STAY COOL

A DESIGN GUIDE FOR THE BUILT ENVIRONMENT IN HOT CLIMATES

HOLGER KOCH-NIELSEN

Published by James & James (Science Publishers) Ltd
8–12 Camden High Street, London, NW1 0JH, UK

A catalogue record for this book is available from the British Library.

ISBN 1 902916 29 8

Printed in the UK by The Cromwell Press

Acknowledgements

This Guide is based on a report produced in late 1999 for the Danish Foreign Ministry, Danish International Development Assistance (DANIDA). The report was used for the training of architects and engineers in countries that receive development aid from Denmark in the energy sector. The consultancy firm Development Advisory Group, of which I am the director, produced the report based on a selection of available literature on the subject and on my own personal experiences, gained when working in these hot dry and warm humid climatic regions. The report was produced with research assistance from Katarina Vrdoljak and Rene d'Amours.

This Guide is a totally revised version of the report. I am grateful to a large number of colleagues who encouraged me in this endeavour and also to those who kindly gave me access to their collections of photographs.

I am grateful to Maurice Mitchell and André Viljoen from the University of North London, School of Architecture, and to Gregers Reimann from the Technical University of Denmark. All three took the time to read my manuscript and give me their comments before the Guide was finalised. Also to Development Workshop, France, for their assistance in the early stages of the production process and to Claire Norton in particular, who kindly assisted me throughout the production of this book. And, finally, to my family for their patience. My thanks are also due to The Danish Arts Foundation and the Plum Foundation, Denmark, who have supported the production of this book.

Holger Koch-Nielsen

Contents

Foreword

By José Forjaz, Director of the School of Architecture as well as an architectural practice in Maputo, Mozambique.

This is a useful book. It is a useful tool for us who teach and practise architecture. It is useful for our students, because it is both clear and comprehensive.

It is useful for the concerned client, keen to make use of natural forces and phenomena instead of turning to imported and unnecessary energy to ventilate and cool his building.

It is useful because it demystifies the quasi-esoteric aura with which some authors surround this subject, which should instead, as this book proves, be brought down to a simple matter of understanding the logic of nature.

My hope is that, some day in the near future, books like this will be mandatory components of the secondary school general knowledge syllabus.

Only then can we hope for human society to demand the necessary level of awareness, competence and commitment to the creation of a sustainable human habitat from the architect and from the planner.

If the social needs, the technical level of sophistication, the economic constraints and the cultural background of an architectural programme are essential to the definition of the built environment, we, at the beginning of the new millennium, can no longer avoid regarding environmental parameters and the sustainability dimension as essential determinants of the form of our cities and architecture.

It is in helping us in this difficult, creative task that this book makes an even more valuable contribution.

José Forjaz
March 2002

Foreword

by Jørgen S. Nørgård, Associate Professor, Department of Civil Engineering, Technical University of Denmark.

Building design has traditionally called for architects. This has often taken the form of what the author terms passive design, leaving engineers to provide a good indoor climate by the use of active energy-consuming measures with their associated environmental problems. If the energy-efficient design of buildings in warm climates could be undertaken entirely by architects with the help of this book, that would be fine. But energy-efficient design also calls for more technical knowledge, and another type of engineer working on the principle that 'prevention is better than cure' could play an important and active role. Cooperation between such engineers and architects could produce energy-efficient, preventive and truly exceptional designs, which reduce or even eliminate some of the cures of complex and energy-demanding technologies. Architectural art and engineering rationality should from the very first outline of a building be united in a team or even in the same person, in order to achieve an optimal design with respect to indoor comfort and environmental sustainability.

It is encouraging that this book does not present final solutions but rather outlines some guiding principles, leaving room not only for climatic differences but also for preserving as well as drawing on cultural diversities, with all the benefits this implies. A conventional engineering approach to buildings has been to make the indoor climate and the use of the building totally independent of nature in a way that fails to stimulate the human mind and body, leaving both unresponsive. In contrast to this approach, this book suggests building designs that involve 'listening' to nature in providing a comfortable and healthy varying indoor climate in an environmentally sustainable way.

Based on energy-efficient building design, the initial cost is usually expected to be and is accepted as higher. This, however, need not always be the case, since it saves the installation of the mechanical systems for cooling, etc., which is also reflected in the cold climates in Europe, where houses based on active design and passive means for heating are built. But generally, an extra initial effort ought to be acceptable in the light of the later benefits. These benefits consist not only in lowering of the running costs of the building and the environmental advantages, but also in more independence from energy supply and mechanical maintenance for the user of the building as well as for the country.

It is a relief to see a well-written book explaining, in a way that is understandable to a broad audience, the basic principles of how to stay cool in a hot climate with a minimum of energy use. With this book no one engaged in building houses can use the excuse of lack of knowledge to continue to design energy-greedy buildings in warm climates. Moreover, government administrators as well as contractors should now know what they might reasonably require as standards for the construction of new buildings, as well as for the renovation of existing ones.

Jørgen S. Nørgård
April 2002

The contents of the Guide and its use

The purpose of the Guide is to introduce the basic principles and tools that can be used when rethinking the built environment from an energy perspective. However, the ideas expressed in the drawings and photographs are not the answers to 'how to design'. Each individual site, taken in context, should give these answers. Also, the illustrations are neither 'the best architecture' nor 'the best engineering' solutions but rather provide some ideas to explain the general principles in the use of passive measures to satisfy human comfort requirements for specific climatic characteristics, based on examples of traditional and modern constructions.

The Guide is arranged in sections, progressing from the outline of these characteristics and requirements to how the built environment, external environment, building envelope, material selections, and natural ventilation and cooling methods can be used and shaped to satisfy the human comfort requirements in hot dry and warm humid climatic zones. All sections are interrelated but, for ease of reading, cross-references have been limited, at the cost of some repetition.

The Appendices include: an active design process checklist and summary of available design checking tools; a rehabilitation guide for existing urban, building and external environments; and solar charts.

Introduction

The purpose of this Guide is to serve as an educational tool and handbook for planners, architects, engineers, technicians and other designers working with planning and design of the built environment in hot dry and warm humid climate zones. It does not aim to cover all architectural, technical and analytical aspects of the subject, but to be an inspirational introduction to general design principles for energy efficiency and comfort taken from the urban planning scale through to the detailing of structures that has to be addressed when working in the climatic characteristics of these two climatic zones. The Guide deals with environmental issues in the sense that they form part of any energy-efficient strategy, as reductions in energy consumption will have a positive environmental impact.

A facade illustrating the use of active means – air coolers, to create internal comfort. From Maputo, Mozambique.

Many excellent publications are available on the subject (see Bibliography), but the majority of these focus on a specific subject matter or target a specific audience, e.g. experienced architects, engineers, ecologists or other specialists.

In the following text, energy-efficient design is to be understood as design that minimises energy consumption in buildings by using natural measures to improve comfort conditions. This is different from designs where internal comfort is achieved by mechanical cooling and ventilation equipment that uses imported energy for its operation. For such a typical air-conditioned building in a hot climatic zone 70–80% of its lifetime total energy consumption is used just for operating these systems for internal comfort. The remaining 20–30% of energy is consumed in the production of materials, construction and demolition of the building.

Why energy-efficient solutions?

At present 83% of the global primary energy produced is consumed by only 25% of the world's population, leaving only 17% at the disposal of the remaining 75% of the world's population (who live mainly in the developing world).

The unsustainability of this situation becomes more acutely apparent when analysed in relation to the global trend towards urbanisation and industrialisation. Today half of the industrialised world's population lives in urban areas and accounts for half of the world's total energy consumption. Only 10–20% of the population of the developing world lives in urban areas. It is predicted that within a few decades 60–70% of the total world population will live in urban areas. Consequently, the demand for energy resources could be enormous.

For many of the countries that fall within the two climatic zones covered by this Guide, half or more of their urban peak load of energy consumption is used to satisfy air-conditioning demands alone. Since it takes more energy to cool than to heat, the pressure placed on energy resources to satisfy the future demands in the hot and warm developing countries will be great unless new measures are introduced.

Although traditional energy resources are sufficiently large for several decades to come, the increase in demand will have considerable environmental consequences. It also raises questions such as who will be able to access these resources. This is particularly important for developing countries, where development budgets should not be exhausted on needless investments in cooling equipment and energy supply.

The relatively cheap energy prices in the industrialised world and the subsidised energy prices in developing countries have provided no incentive for clients to pursue energy-efficient design options. The current and growing concerns regarding the impact that traditional fuel-based energy consumption has on the environment has spurred the move to find alternatives and to examine the use of resources based on their life cycle and real costs. It is likely that energy prices will be increased in the future in order to deal with pollution and the growing concerns for the depletion of traditional resources and/or the use of more modern non-polluting energy resources.

Energy, particularly from traditional sources such as coal or crude oil, cannot be supplied without the use of a considerable amount of energy. The total efficiency of the system from the extraction of resources, preparation, transport, conversion, and distribution to application may only be up to 20%: that is, it takes approximately 5 kWh to deliver 1 kWh of energy. Consequently, the energy saved by the end user has a greater impact on energy resource demands, energy investments and the environment than any other measures.

Alternative energy sources (e.g. solar power) may in the future be more feasible and in common use. However, there are no concrete indications (i.e. based on real costs) that alternative energy sources can be produced at levels to satisfy fully current energy demands. In developing countries, if future energy demands are to be sustainable and affordable, alternative approaches that promote energy efficiency need to be embraced.

Potential savings can be gained by activating specific, energy-efficient technologies, but more so through careful design of urban environments and individual buildings: all naturally occurring

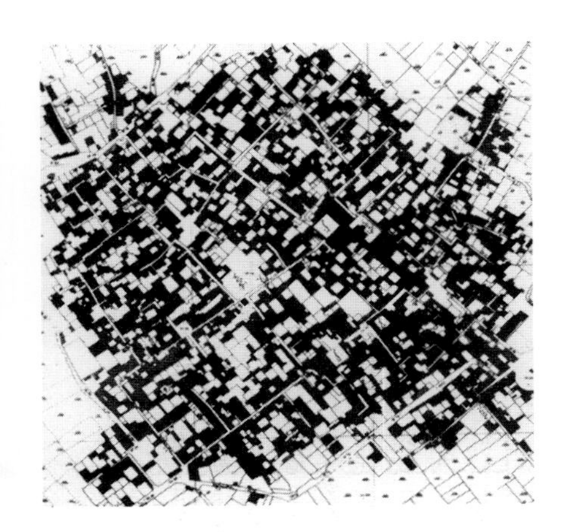

In hot dry regions the built environment should be planned compactly to reduce the amount of surfaces exposed to solar radiation. An aerial photo of an unknown city in Omar.

In warm humid regions buildings should be openly spaced to maximise air movement between individual buildings. The drawing shows a possible urban plan for warm humid regions.

resources should be integrated into planning and building and in such a way that their location, form and structure promote energy savings.

Thinking actively

Designing in an active spirit was introduced to me in the writings of John Berry, Director of Ove Arup, Consulting Engineers, London, and his spirit has been adopted by this Guide and translated to active design. The term 'active design' is to be understood as developing design solutions that exploit passive measures (e.g. building envelope design, climatic conditions and natural energy sources) to achieve the desired indoor and outdoor comfort conditions. This is contrary to notions of passive design, where internal comfort is achieved primarily through active measures, i.e. cooling and ventilation based on importing energy into a building. It is not always possible to avoid active measures, for instance due to local conditions or under extreme climatic circumstances. A mixed approach of both active and passive design may be necessary, especially for buildings that demand a high, constant level of internal comfort.

Active design for energy efficiency of the urban, built environment requires an integrated approach involving all disciplines – urban planning, architecture, engineering, construction, etc. – at all levels of the design process and through all details of a building. The design must also include operation and management of the buildings and the development of adaptation, reuse or demolition strategies as well as the use of local materials and construction processes that embody energy-efficient approaches, though this subject falls beyond the scope of this Guide.

There are differences in the management of building systems depending on whether they are based on active or passive measures. Active measures involve mechanical equipment, which requires constant mechanical regulation. The advantage of such a system to many developers and clients is the possibility of fully controlling the internal conditions, regardless of daily or seasonal variations in the external climate. Once inside the building, nature can more or less be neglected or forgotten. The disadvantage of such a system is that it demands constant maintenance and careful operation, and that it has high and ongoing running costs. If an unreliable power supply is in place, the added expense of a back-up system will also be required.

Passive measures rely on utilising the elements inherent in a region's climate and its natural energy sources. This involves 'listening' to nature and reacting to the changes that occur with the seasons of the year and the time of day, e.g. closing and opening windows according to changes in temperature, prevailing wind conditions and adjusting natural cooling systems. To use passive measures is to accept the dynamics of nature. This often means that a total and constant control of a building's internal climate may not always be possible. Likewise, it is not always guaranteed that a building that relies solely on air conditioning will be able to achieve appropriate internal comfort conditions or indeed an environment that is healthy.

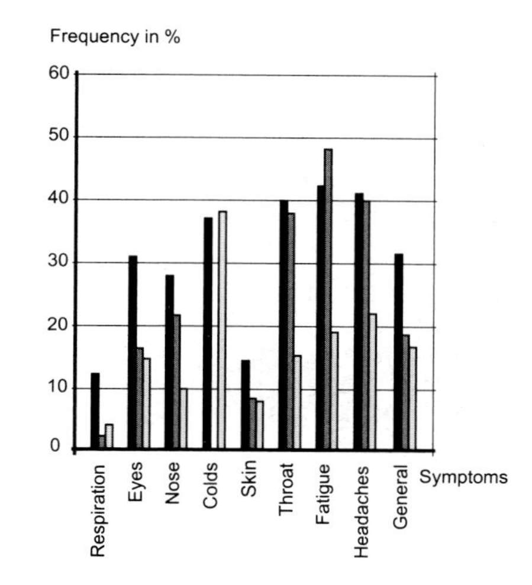

Diagram indicating illness frequency in different types of building. Based on a study by J. Roben mentioned in The Technology of Ecological Buildings *by Klaus Daniels.*

A move towards passive measures based on active design

In the past, buildings were constructed using passive measures, quite obviously due to the lack of any energy supply or resources. In industrialised countries, active measures gained predominant use during the post World War II construction boom, as an easy and quick means of satisfying comfort concerns in buildings.

In this process architecture lost its connection to 'place'. It was now possible to create built environments that were totally controlled and divorced from their surroundings. This may consequently have resulted in the lack of attention paid to the real costs associated with this form of progress.

It was not until the energy crisis in the 1970s that the quantity of energy consumed to maintain these new environments was seriously addressed, particularly in the cold climates of Europe.

Climate-responsive building design. In hot dry regions the building envelope should employ dense materials and few openings to reduce heat gains. Local materials, such as adobe wall constuction, have proven ideal climate-responsive solutions. A building from Cairo, Egypt.

In warm humid regions buildings should be light and transparent to air movement. In the above example, walls are made of a light timber construction, with air gaps between each slat to maximise continuous air movement and cross ventilation of internal spaces. A building from the southern parts of Burma.

A sense of place: working in context

In many developing countries passive measures have been equated with tradition, and these may not always conform to the images of modernity. This may account for the general lack of acceptance, which is understandable, as a similar belief drove the development of the industrialised world.

Some countries, particularly in the hot dry climate zone, developed urban areas before any such forms existed in Europe. Buildings constructed from adobe or burnt brick had compactness in planning and form that addressed their context. Traces of these actively designed buildings and urban areas celebrating their sense of place are still to be found in Africa, the Middle East and Latin America.

Wide streets, laid out in grids, have replaced this generally unsustained tradition with rows of individual houses on individual plots of land. This model, a significant trend of European city planning throughout the last centuries, is today found in most cities in hot climates.

The climate and landscape inherent to a region has through time served as the foundation for the design of the urban and built environment. Appreciating and using nature as a source of inspiration should not be confused with a lack of development or desire for progress. To design according to the characteristics present in nature will lead rather to an architectural vocabulary that is shaped by rational, reasoned and proven solutions. It also produces a language that is accessible to all, and its inherent attractiveness is 'natural' owing to a harmony with its environment.

Contemporary urban or architectural interventions in areas that are situated in the climatic zones covered by the Guide have often been based on models imported (or imposed) from areas that climatically have little in common. There is an abundance of building examples that make little or no reference to their context. The use of inappropriate building forms, materials and openings, and the failure to provide protection from the fundamental cause of discomfort such as solar radiation, indicate this clearly.

Preconditions for a successful introduction of active design

Short-term thinking, which focuses on the possible added construction costs and the need for more time and fees in the design development phase of a building, works against active design/passive measures. This is regardless of the fact that the long-term running costs and the environmental impact of active measures are more expensive, and that their environmental consequences are known.

Introducing new ways of conceiving the urban and built environment in the countries covered by the Guide is not possible without full support from all parties involved. A mixture of motivation and education has to be introduced. This involves the education of authorities, clients and designers, and a change in attitude in which energy conservation is given priority, and measures to achieve it are adopted at all levels.

Authorities should adopt urban planning principles and energy-efficient design criteria similar to those that are outlined in this Guide. Clients and developers can be motivated by government initiatives such as campaigns that raise environmental consciousness, higher kWh rates for electricity consumption for environmental reasons and government regulations and codes concerning energy consumption, e.g. a fixed maximum kWh/m^2 per annum for different building types.

Clients and developers should be made aware of the advantages offered by energy-efficient buildings. The clients also have to accept that the building is not finished until its users have been trained to coexist actively with their surroundings, i.e. have learned to operate the ventilation system and natural cooling system according to the time of day/year. A developer, understood as the person who finances and manages the construction of new buildings and who does not occupy the building himself, is, however, often a difficult client for an active design team as his main concern is not the operation costs of the buildings, but merely the initial construction costs. This situation, in particular,

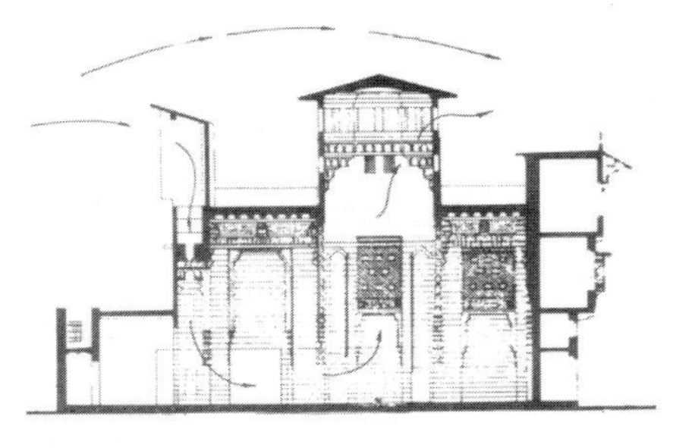

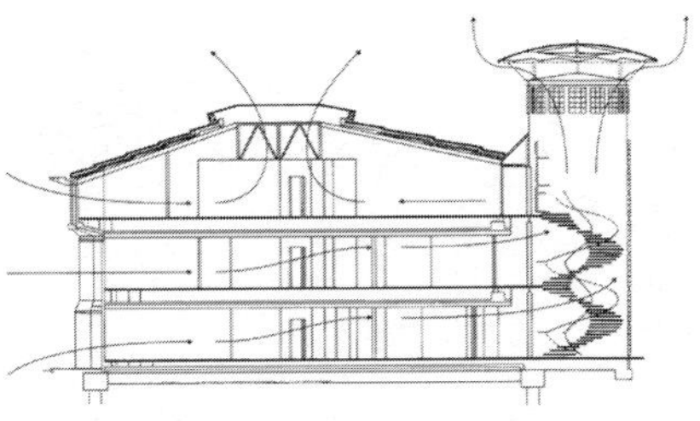

Energy-efficient design solutions.
(top) Local and time-tested solutions can be incorporated into buildings. The traditional windcatcher draws on natural wind energy as the motor to cool internal spaces. Building in Cairo, Egypt.

(above) A contemporary solution using the stairwell as a ventilation device. In this case the cooling and ventilation of internal spaces is achieved by naturally occurring temperature and pressure differences. A building in Nottingham, UK.

makes thermal building codes necessary in order to achieve energy efficiency and energy conservation in the building sector.

It is the designer's responsibility to make the client aware of the means and advantages of lowering the energy consumption of a building. In this respect the Guide aims to assist designers in the development of active design strategies. It is important that designers are always commissioned to develop a building's own energy manual and provide guidelines for its enforcement during the operation and occupation of the building.

Codes and rules in developing countries should be based on the conditions in the specific country and not on foreign standards.

The approach: how to design actively

Architects generally work in a process-orientated manner, i.e. examination of the brief and its functional requirements, analysis of the site and climate, the development of a design concept and the documentation and detailing of the design, etc. until all design criteria are met. This process involves backtracking and revising previous decisions, a process that is cyclical rather than linear in nature.

In a passive design situation the architect generally designs a building, and the engineer will afterwards try to adapt the building to its surrounding climatic conditions. In most cases this entails the use of active measures (e.g. mechanical ventilation and/or air conditioning) in all spaces without much concern for energy consumption.

Close professional cooperation between architects and engineers is a prerequisite for the implementation of active design. This is due to the complex nature of applying active design solutions or a mixture of active and passive design solutions to satisfy the functional requirements of contemporary buildings.

In an active design process, it is imperative that thermal evaluations of an architect's designs are made throughout the design process. The passive measures employed in the design of a building to achieve energy-efficient solutions will have a direct influence on the architecture. A combination of active and passive measures may also be needed as a compromise for difficult sites or for other reasons.

The engineer should carry out both static and dynamic analyses. Computer modelling can be used for this, but it cannot yet cope with all the dynamics of nature. Simple analyses regarding the effects of several combinations of natural phenomena have to be carried out. Physical models are also useful tools for evaluating internal heat distribution, air flow patterns and lighting requirements.

Active design also relies on the thermal properties of a building's envelope to create a protective barrier against the external environment. For instance, in hot dry climates a building envelope with high external reflective properties (to minimise the impact of solar radiation) and high thermal storage capacity (to balance the variations in temperatures) is desirable from an active design perspective. In addition, controlled ventilation (i.e. ventilating only during the night) and natural cooling methods must be used to provide additional possibilities in maintaining the desired internal comfort conditions. Openings in the building envelope for daylight are a critical issue in an active design process in order to serve contradictory principles – to avoid heat gains coming through the openings and at the same time to minimise energy consumption from artificial lighting.

There are examples in this Guide from contemporary buildings designed for hot climates that incorporate active design principles. Also included are traditional, active designs based on collective experience, knowledge of the land and climatic conditions. The aim of the Guide is to introduce principles and a variation of options for active design in the two climatic zones concerned.

A good active design solution can be developed only for a specific site with due respect to the context, which in this Guide is the aspects of physical nature. However, the context must be understood in its global sense and include the financial, economic, and social situation of 'the site'.

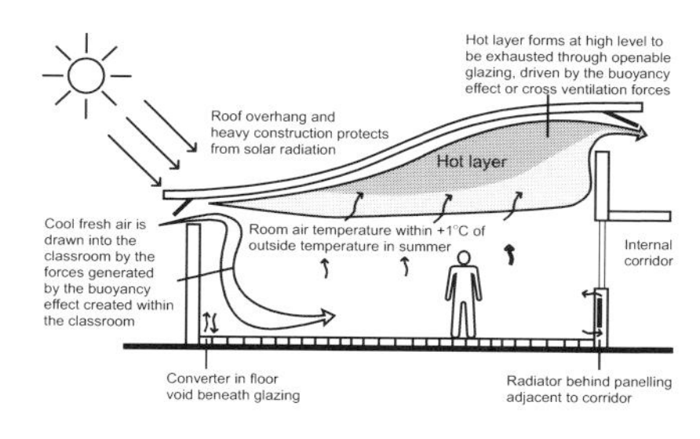

Active design requires that decisions made throughout the design process be rechecked and changes made accordingly. This example is from a building in Coventry, UK.

(top) Sketch design illustration showing the climate-responsive decisions made early on in the design process.

(middle) The effectiveness of these solutions can then be verified using various design-checking tools. In this case a saline test model is used to demonstrate the heat gains from internal heat sources.

(above) The finished building.

I. Climatic issues

The climate of a given region is generally defined as the interaction of several meteorological elements. The principal elements are:

- *solar radiation*
- *air temperature*
- *humidity*
- *wind*
- *precipitation.*

These data should be collected and analysed as the first step in understanding the overall climatic conditions of a particular area. They are normally obtainable and must be analysed on the basis of several years of data in order to establish the design-climatic conditions (design year or period). Site-specific climatic elements, which are normally not readily available, must be observed and analysed together with the regional data.

The following features, which also influence the climate of the specific site, should also be examined:

- *topography*
- *ground cover and vegetation*
- *water*
- *building densities.*

Generalised climatic zones have been developed to determine common design responses required for the different zones. The characteristics used to define hot dry and warm humid climatic zones are provided at the end of this chapter. This information is the basis for the design responses described throughout this Guide.

Solar radiation

The intensity of solar radiation experienced by an area depends ultimately on the altitude (i.e. the position) of the sun and the quantity of particles in the atmosphere. It will vary greatly, depending on the geographical location and local weather conditions.

Essentially, solar radiation will be affected by the length of day, the angle of the sun's rays to the ground, the distance from the sun, the cloud coverage and the quality of the atmosphere through which the radiation passes.

Radiation is experienced in the form of wavelengths. There are two distinct types of wavelength that will affect the heat gains and losses of a building:

- short-wave radiation; this is also referred to as high-temperature or visible radiation as it is radiation received from the sun or other very high-temperature sources; and
- long-wave radiation, which is also referred to as heat or invisible radiation, as it is the radiation that is emitted from a hot object such as a building surface and absorbed by less hot objects or the atmosphere.

There are three main types of radiation that will affect a building and the amount of heat transferred into a building:

- direct solar radiation, which is the component of radiation received directly from the sun (from the direction of the solar rays);
- diffuse solar radiation, which is the component of radiation that is received from the whole sky vault, i.e. the sky, clouds and atmosphere (from all directions); and
- reflected solar radiation, which is the component of radiation that is reflected from the ground and other objects, e.g. other buildings.

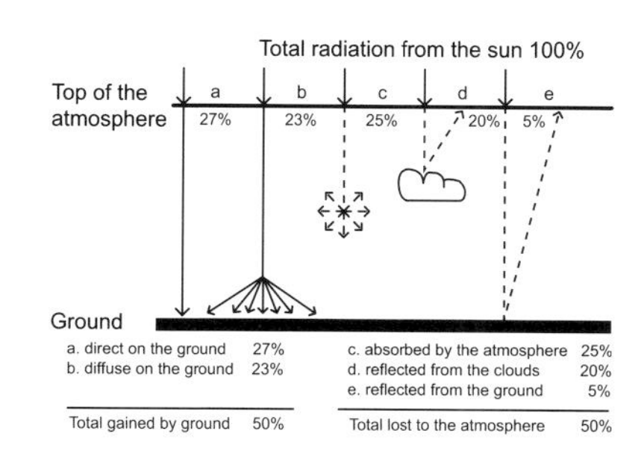

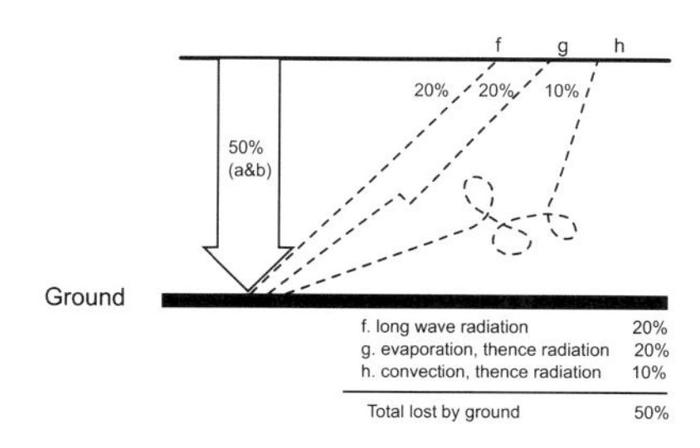

The passage of solar radiation through the atmosphere and its impact on the environment. The distributions of the components in the diagrams are average figures. They will in general vary according to the weather conditions and the time of day. The diagram below shows the radiation heat gains and losses that take place on the ground. Gains normally take place during the day and losses during the night. If the balance between gains and losses is disturbed, the ground temperature will change. The greenhouse effect will for example reduce the radiation losses, which will cause higher average temperatures on the earth's surface.

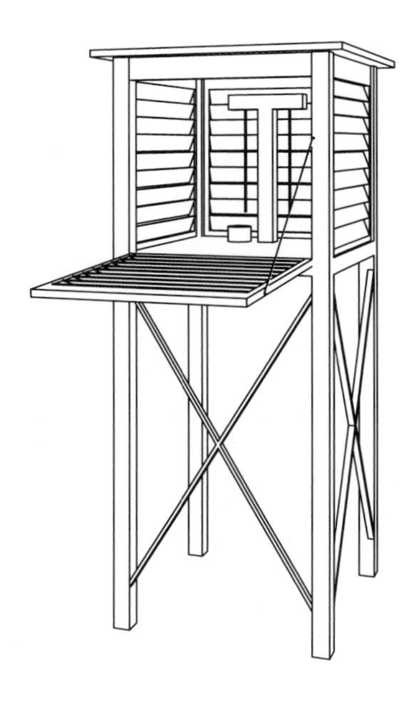

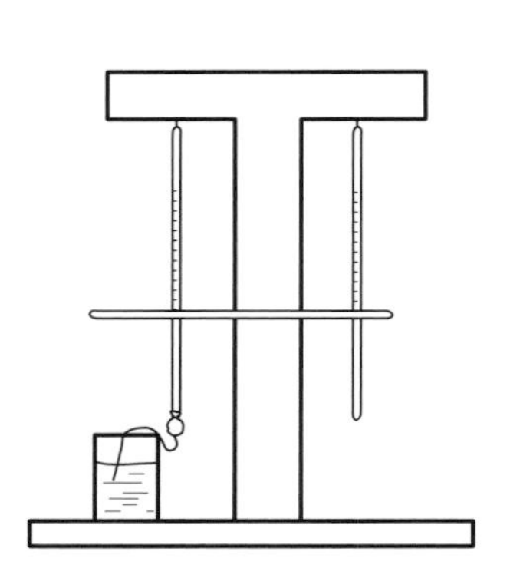

(top) Shows a classical Stevenson cage for meteorological temperature measurements. (above) A close-up view of the wet bulb (wetted) thermometer and the dry bulb thermometer.

Air temperature

Air temperature is in meteorological terms a measurement taken in the shade approximately 1.5 m above the ground in a ventilated box, in order to prevent radiation from the sun and the ground from affecting readings. Air temperature is normally referred to as the dry-bulb temperature as opposed to the wet-bulb temperature. Both temperatures are measured simultaneously in the same ventilated box using two thermometers. The wet-bulb temperature is measured simply by keeping the thermometer humid in a wet piece of cloth. With the two temperature measurements, the specific humidity and the relative humidity can also be determined. Meteorological air temperature is not the same as the temperature or heat experienced in the built environment by humans, where solar radiation and radiation from the environment affect the experience of temperature.

The degree of heating and cooling of the surface of the earth is the main factor determining the temperature of the air. Generally, temperatures are lowest just before sunrise and are highest two hours after noon, when the effects of direct solar radiation and high air temperatures are combined.

It is important to try to obtain not only the maximum temperature, but also the minimum temperature variation for a region. This will give an indication of the diurnal variations (i.e. the temperature difference between day and night). Since a large diurnal range is indicative of dry weather and clear skies, one can anticipate intense solar radiation by day and strong outgoing radiation by night, whereas a small diurnal range indicates an overcast sky and a high humidity range.

Humidity

This term refers to the water vapour content of the atmosphere gained as a result of evaporation from exposed water surfaces, from moist ground and from plant transpiration. For any given temperature there is a limit to the amount of water that can be held as vapour, and the air's capacity increases progressively with increases in temperature.

Terms used to describe the moisture content in air relevant from a design point of view are as follows.

- Relative humidity is defined as the ratio of the actual humidity in a given volume of air to the maximum moisture capacity at that particular temperature. It is the term most commonly used to describe humidity.
- Vapour pressure is the part of the total atmospheric pressure that is due solely to the water vapour. The term is used in meteorological data and in technical literature.

Wind

Direction, speed, gustiness and frequency are the most important characteristics of wind. Wind roses are normally used to indicate the characteristics of wind for a specific period of time, e.g. a day, a month or a year. Depending on the origin of a wind its quality will differ. It can be dry or humid, clean or dusty, hot or cool. Wind is a very unstable parameter as characteristics fluctuate according to prevailing weather conditions.

The winds over a region, their distribution and characteristics are affected by both global and local factors. However, the principal determinants are seasonal differences and daily variations. In West Africa in the winter, for example, hot dusty winds are from the north-east, and in the summer humid winds are from the south-west.

The direction, speed and predictable daily and seasonal shifts of prevailing winds must be determined and analysed so that their positive or negative aspects can be used or overcome.

There is a variation in wind speeds depending on the terrain over which the wind blows, e.g. urban areas and wooded country offer protection to the open flat country. The terrain will not only affect its speed, but also its quality (hot or cold winds, dusty or clean winds).

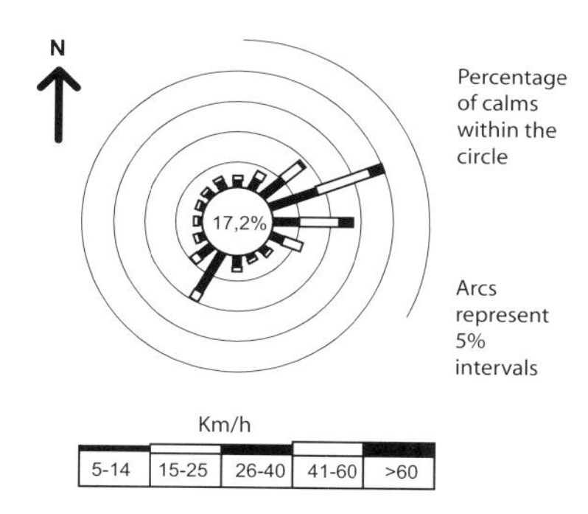

To illustrate wind characteristics a wind rose is often used. This arbitrary wind rose could represent data collected for a period of a week, a month or a year. For natural ventilation design the wind characteristics distribution during the 'design hours' has to be established, data that are not always readily available.

Precipitation

It is important not only to ascertain the total rainfall for each month of the year, but also the maximum amount for any 24-hour

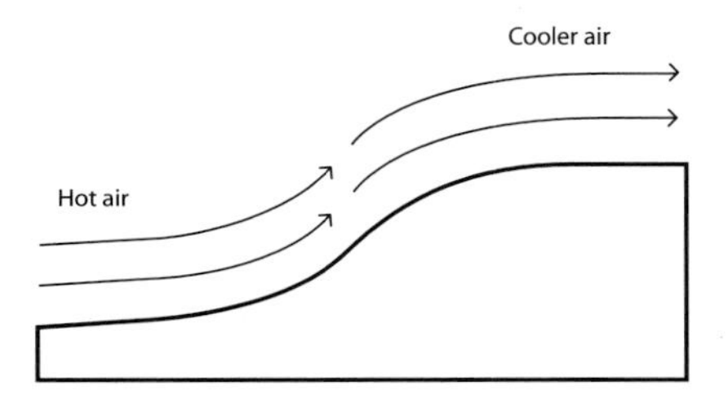

Temperatures will vary with altitude.

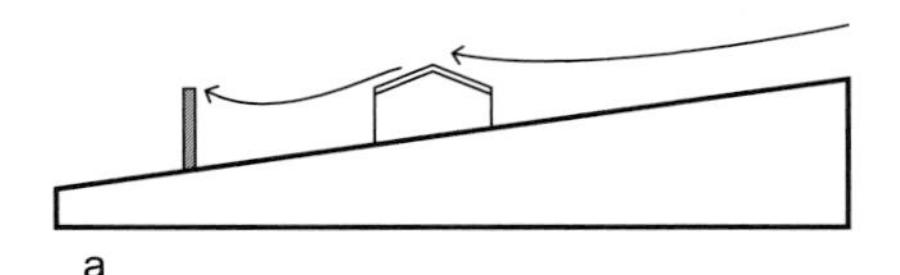

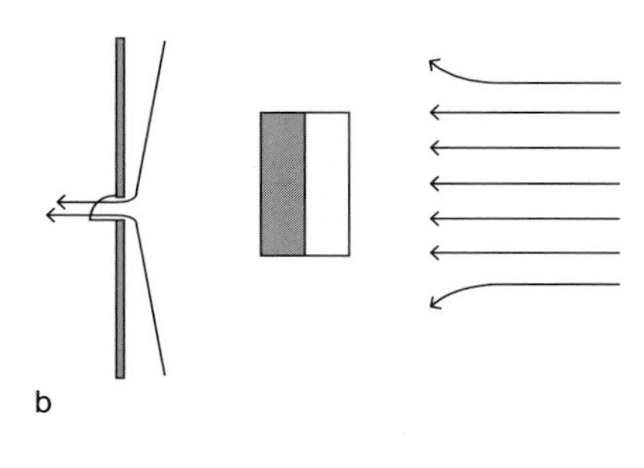

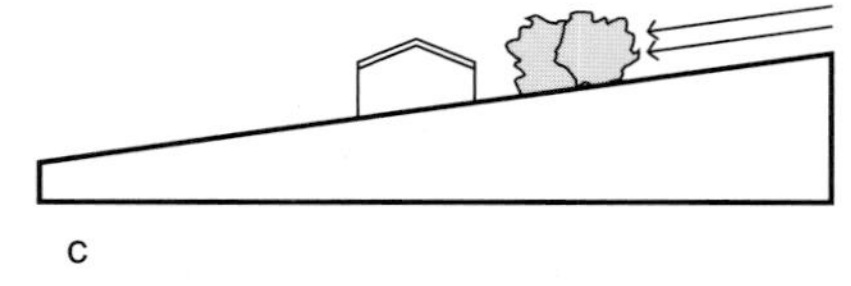

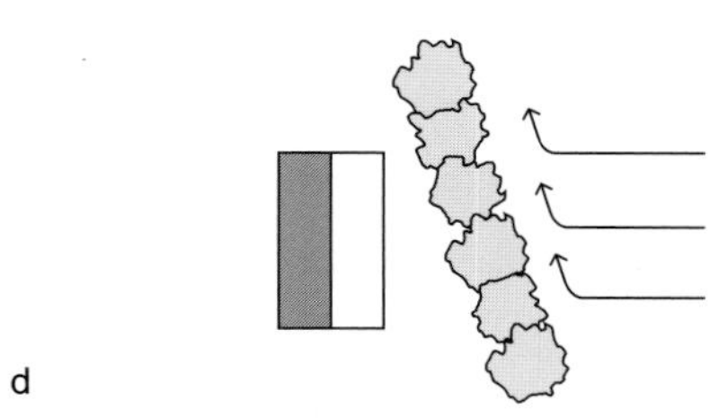

At night cold air drains to the lowest point. a. It can be damned by a wall (or hedges). b. The wall can have an opening to stop cold air build-up in cool periods. c & d. A diagonal barrier (e.g. planting) can be used to protect buildings from undesirable breezes.

period so as to ensure adequate drainage from roofs and paved areas. It must also be determined whether there is a likelihood of driving rain, and whether there are unusual climatic phenomena such as hail, hurricanes or tornadoes.

Features that will influence the site climate

Topography

The altitude, slope and exposure of the site should be investigated, as these will have an effect on solar radiation gains, air temperatures, winds and precipitation.

Altitude

Solar radiation intensities increase with increase in altitude. This is because solar radiation passes through a thinner layer of the earth's atmosphere before reaching the ground. At night, the opposite occurs, as the ground will lose more heat by radiation to a clear sky. This leads to greater temperature variations at higher altitudes.

Air temperatures decrease with altitude. For example, a distance of 7–8 m from the ground may produce a difference of up to 5°C in air temperature under still conditions due mainly to the reduced impact of ground radiation. This decrease in temperature is not incremental and is normally measured at 0.5–1°C for each 100 m above ground.

At night this effect is reversed as cold air drains down to the lowest points. In hot regions this factor can be used to benefit cooling, as a raised embankment, a wall or hedges on the lower slope of a site can dam the cool air moving slowly down the slope.

Slope

When the sun shines on a sloping surface the variation in direct solar radiation will be greater than on a horizontal surface, i.e. flat terrain. Radiation will depend on the altitude of the sun and the orientation of the slope to the position of the sun. Therefore, when radiation falls on slopes facing in different directions, the heating effect will vary.

Generally, slopes greater than 15° will have an effect on site climate, e.g. prevailing winds and air temperature. East-facing slopes will receive a maximum intensity of radiation before noon when air temperatures tend to be low, while west-facing slopes will experience a maximum intensity in the afternoon when air temperatures are high.

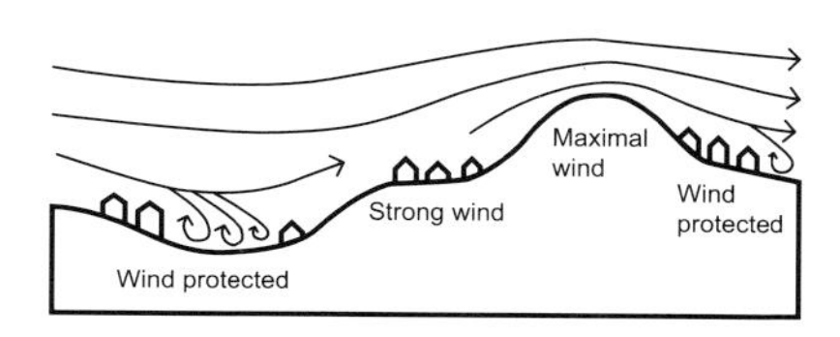

Strong winds are experienced on the windward side of hills; winds will be diverted by hills; valleys may funnel winds and generate their own thermal winds as air is drained down slopes.

Exposure

Topography also has an influence on wind characteristics. Valley bottoms are generally wind-protected areas whereas elevated locations are exposed to more and stronger winds.

Topographical maps of the site and its surroundings must be examined to avoid building in flood-prone areas.

Ground cover and vegetation

During the day the highest temperatures are found just above the ground. At night, because of evaporation and the effect of outgoing radiation, the reverse is true, and the temperature is lower closer to the ground. That is, the closer to the ground, the more extreme the conditions.

The natural cover of a terrain tends to moderate extremes in temperature and stabilise conditions. Plant and grass cover decrease temperatures, but buildings and artificial structures and pavements will absorb heat from solar radiation and increase surface temperatures, e.g. asphalt absorbs almost all the solar radiation and concrete absorbs about 50%.

Vegetation will influence solar radiation gains, humidity levels, wind speeds and directions and will allow a certain amount of air to pass through, causing less turbulence than solid structures.

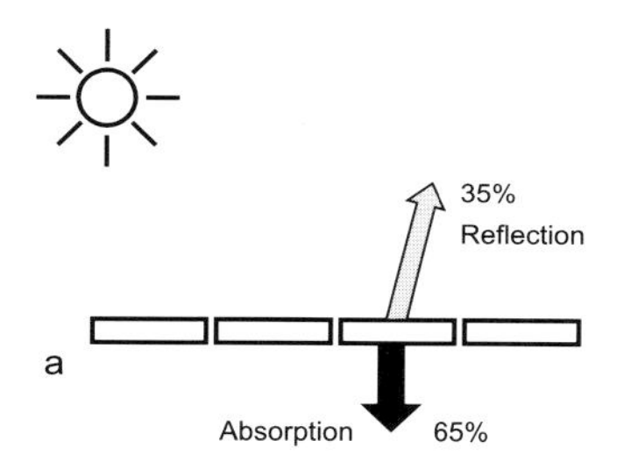

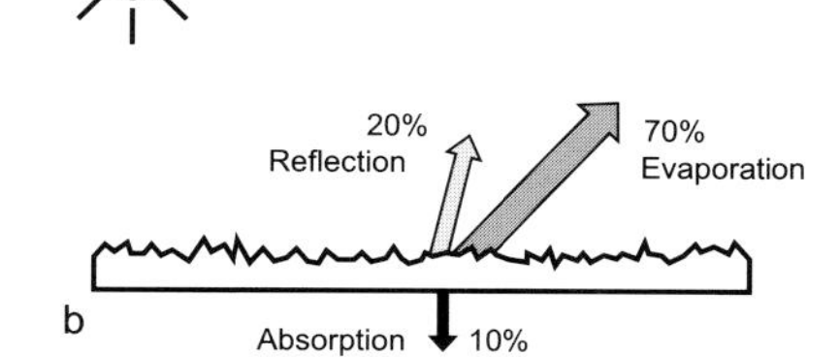

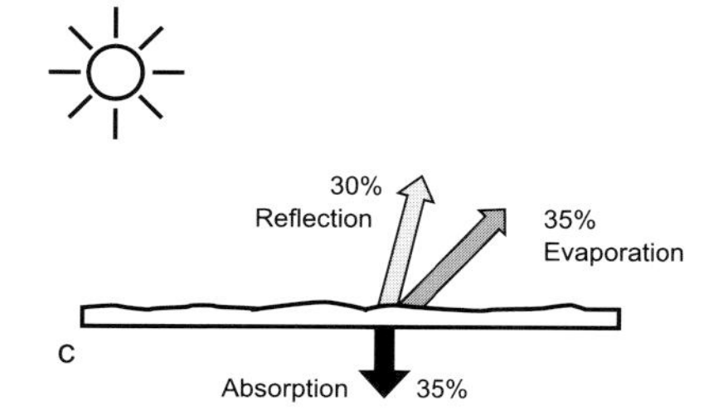

Ground cover variations will impact on the amount of solar radiation that is either gained or lost: a. paving, b. grass, c. bare ground.

Water

Since water has about four times the thermal storage capacity of dry land, it will experience a lower range of temperature variations, both during a single day and over the year. Even regions

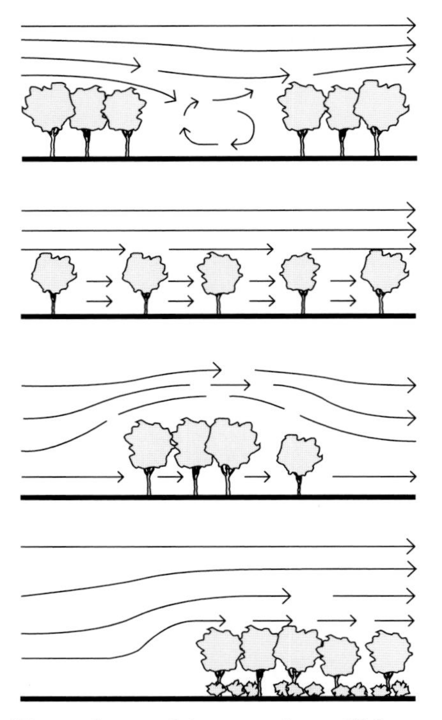

Vegetation and its grouping will have an effect on wind movement.

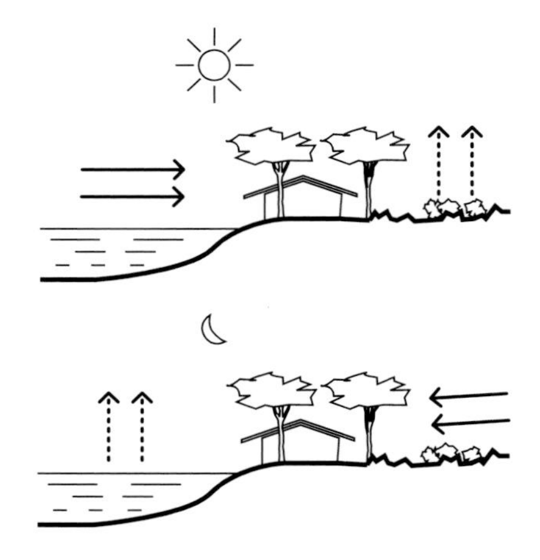

The direction of breezes between land and water will vary. During the day on-shore breezes will occur and during the night this is reversed to an offshore breeze.

10 km away from the sea can benefit from a reduced range of temperature variations.

The difference between coastal and inland locations will tend to be much greater in dry climates, which accentuate the differences in humidity and diurnal temperature ranges.

During the day, solar radiation heats up the land more than the sea, and this creates on-shore breezes. As the hotter air over the land rises, the cooler air from the sea flows inland to replace it. During the night, this process is reversed as the land cools faster than the sea and creates offshore breezes. Sea breezes can be felt up to 3 km from the coast, but topography, planting and buildings may obstruct their effects.

Building densities

In urban areas, buildings sited close together will shade each other, which may reduce the heat gain of a shaded building. But buildings and paved areas will also store and reflect heat, which will increase their ambient temperature. In big metropolitan cities this phenomenon, together with the heat generated from human activities, mechanical and electrical equipment, transport, etc., produces what is called the 'heat island effect'. Studies in Athens in Greece indicate (from *Natural Ventilation in Buildings* by Francis Allard (Ed.)) an extreme heat island effect, where the temperature in the urban town centre has been measured to be 5–17°C higher than in the suburbs. In the National Garden, a completely green environment situated in the urban area of Athens, measurements show temperatures up to 10°C lower than those of the urban town centre on hot summer days. B. Givoni mentions in *Climate Considerations in Building and Urban Design* interesting studies of high-density and tall buildings which create a below and above canopy (roof level) microclimate where the radiation is unable to reach the ground, thus affecting heat gain and heat loss. Tall buildings are not, however, the subject of this Guide.

The two climatic zones.

Definition of the climate zones

Hot dry zones

Hot dry climate zones are normally found between 15° and 30° north and south of the equator. Areas closer to the equator may experience a short humid period, and those further away from the equator may have short cool periods. The main characteristics of hot dry zones are as follows.

Solar radiation

The solar radiation is direct and strong during the day. The absence of cloud cover allows the heat in the form of long-wave radiation to be released easily from the ground or structures towards the cool night sky.

Air temperature

There are mean maximum day temperatures in the shade of around 45°C during the hot season (temperatures up to 50°C in the shade may occur) and between 20°C and 30°C in the cold season; mean night-time temperatures of around 25°C during the hot season and between 10°C and 20°C in the cold season. In desert regions, however, night temperatures can drop below 0°C in the cold season. A diurnal temperature range of 20°C is common. The annual temperature range depends on latitude. With increasing latitude the annual range is greater.

Humidity

Average relative humidity is between 30% and 40%, though there may be increases in the cool season. Potential evaporation may be high (so the skin will remain dry). Maritime areas experience higher relative humidity levels of 50–90%, which offshore breezes will only partly counteract.

Rainfall

Average annual rainfall is extremely low, ranging from 50 to 500 mm. Rainfall may be variable and unreliable. Averages mean very little as there may be long periods without rain, while one sudden storm may bring rainfall exceeding the annual average.

Sky

This is generally clear and deep blue, so sunshine is abundant, resulting in strong surface heating due to intense solar radiation. These conditions predominate, with occasional light cloud cover in the early morning and in the cool season. However, after sandstorms small particles may remain in the sky, creating a haze and painfully bright skies. This increase in atmospheric particles will create an artificial sky, leading to an increase in diffuse radiation.

Winds

These are usually local. The heating of air above the hot ground may cause temperature inversions, with warmer air masses breaking through higher cooler air, creating local whirlwinds. Winds will tend to increase in intensity from dawn through to the afternoon and drop at sunset. This may raise dust and create a dust haze in the afternoon. Occasionally strong winds will cause dust storms.

Groundcover

The ground is dry and barren with red/brown sand and rock. Natural vegetation is sparse or non-existent. This lack of vegetation results in high winds at low levels, which may cause wind erosion. Vegetation that is protected from the wind and artificially irrigated will thrive.

Warm humid zones

Warm humid climate zones are normally found in the latitudes close to the equator between 15° north and 15° south. There may be a dry season of less than two months but it is predominantly humid and rainy.

Solar radiation

High humidity and cloud cover reduce direct solar radiation, but increase the proportion of diffuse radiation. Direct radiation is more frequent in the afternoons. Glare from the sky is strong, and reflected radiation from the ground is low.

Air temperature

There are annual mean maximum day temperatures of around 30°C in the shade and an annual mean minimum night temperature around 24°C. Both diurnal and annual mean temperature ranges are quite narrow: diurnal ±8°C, annual ±3°C.

Humidity

This is high throughout the year, with average relative humidity above 60% and commonly around 100%.

Rainfall

This is high, with annual mean rainfall in excess of 1000 mm but with a possibility of one or two dry seasons. Some areas may experience lower rainfall, but still have warm humid conditions throughout the year.

Sky

The sky is fairly cloudy throughout the year, with 50% cloud cover or more. Thin cloud cover or broken cloud can lead to very bright skies and strong glare. Dark, heavily overcast skies often precede heavy rain.

Wind

Average wind speeds are low, but one or two predominant wind directions are usual. Wind direction may vary at different times of the year.

Groundcover

Vegetation grows rapidly, but soils are not particularly productive for agriculture. When vegetation is cleared, organic materials are easily washed out of the soil (i.e. by leaching), and erosion may occur. Damp soil and ground cover reflect little light. High humidity encourages mould and algae growth. Mosquitoes, termites and other insects are abundant and are a source of nuisance.

Notes

Variations will exist within these two broad climatic zones found in the tropics. An area may experience a mixture of the characteristics of both climate zones as well as variations caused by natural features such as water bodies, topography, etc. Some books identify several different climates within the two climatic zones mentioned in this Guide. However, as a general rule if the design principles described for the two main zones in the present Guide are well understood, the complexities in mixed or extreme climates on a specific site within the tropical zone can be addressed.

2. Thermal comfort requirements

The main function of the built environment is to provide protection from the stresses imposed by the climate.

The notion of thermal comfort is difficult to define. In general, it is the sensation of well-being of an individual in a specific environment. Thermal comfort will vary from one individual to another and is therefore subjective. It encompasses not only physiological, but also psychological aspects. The limits for thermal comfort vary with the degree of adaptation to the climatic and the psychological/social environment. Also a person's age, sex, body type and state of health influence the experience of comfort. In the following text, these aspects and conditions are not examined in any detail. Interesting discussions on the diversity of circumstances that affect the experience of well-being are among others found in Naturally Ventilated Buildings *by Derek Clements-Croome (Ed.).*

Since maximum thermal comfort conditions can rarely be achieved, buildings should aim at creating environments within which uncomfortable conditions can be addressed, i.e. limited as much as possible. It is, however, necessary to understand what causes discomfort in order to indicate which particular method of achieving comfort will be most effective.

Human comfort

In order to feel comfortable, humans must maintain an internal body temperature within a narrow range of 36.5–37°C. Metabolic processes that occur in the human body produce heat, and the body must lose this heat in order to maintain a stable internal temperature. Some processes are continuous (e.g. breathing, blood circulation, heart beat) and are called involuntary control mechanisms. Others such as muscular activity, clothing, posture and location are controlled and are called voluntary control mechanisms.

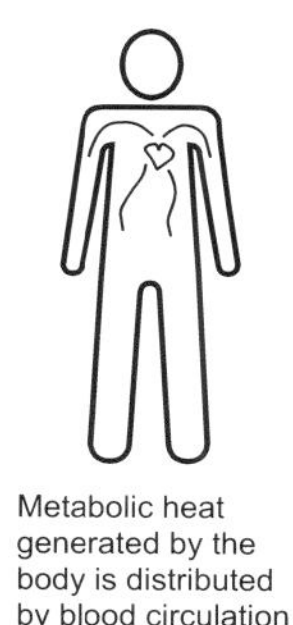

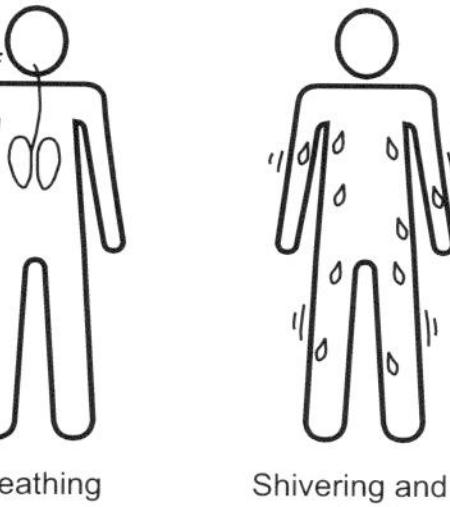
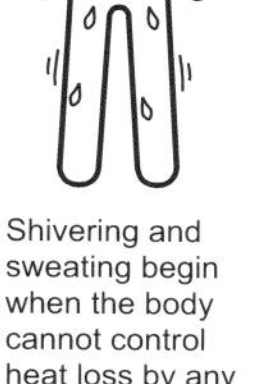
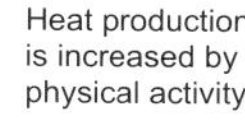

Various involuntary and voluntary thermal control mechanisms of the human body.

Thermal exchange between the human body and its environment

Since the human body aims at maintaining a constant internal temperature by releasing excess heat to its environment, there is as a result a continuous exchange of heat between the body and its surroundings. This exchange occurs in four physically different ways: conduction, convection, long-wave radiation and evaporation – the fraction of each exchange mechanism varies according to the thermal environment.

- Conduction depends on the thermal conductivity of the materials in immediate contact with the skin. This process is limited to the local cooling or heating of particular parts of the body when they come into contact with cold or hot materials. This is of practical importance in the selection of materials and surface finishes that will come into direct contact with the body, e.g. floor coverings.

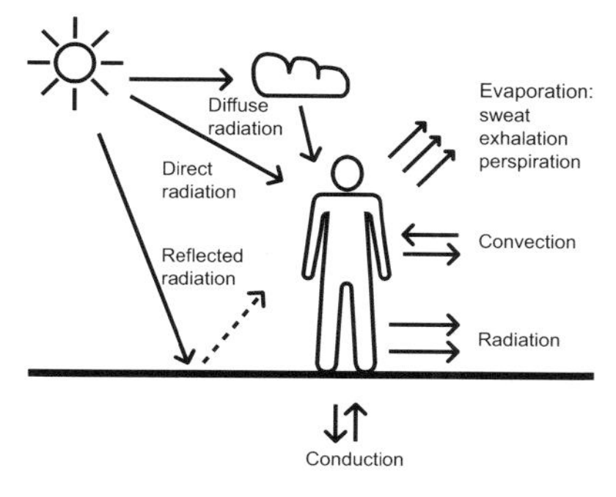

The heat exchange process between the human body and its external environment.

- Convection involves the exchange of heat by the body to the surrounding air. This process will depend primarily on the temperature difference between the skin and the air and the movement of the air. These exchanges are not uniform over the body and are more pronounced at the extremities (e.g. head, hands and feet) owing to the higher sensitivity to temperature variations of these areas. These considerations are important for design decisions such as ceiling heights and the positioning of openings to direct cooler air across the body.
- Inside a building, long-wave radiation takes place between the human body and surrounding internal surfaces (e.g. walls, ceilings and floors). Actual temperature, humidity, and air movement have little influence on the heat transmitted by this process. It depends rather on the differences in temperature between the skin and the surfaces that surround or enclose it. As for convection, it has been found that radiation from/to the extremities of one's body has a higher effect on thermal comfort than for the rest of the body. Tests show clearly that occupants react much more to a hot ceiling than to a hot wall. These considerations are important for design decisions such as ceiling heights and the thermal properties of ceilings and walls.
- Evaporation takes place when the surrounding temperature (i.e. air and surface temperature) is above 25°C. In this case a clothed human body cannot lose enough heat by either convection or radiation, and the loss through evaporation (i.e. perspiration) becomes the primary mechanism. Heat is lost when evaporation takes place, and as humans normally lose around one litre of water per day due to perspiration and respiration, a fair amount of heat is taken from the body to evaporate it. The extent to which heat is lost by evaporation depends on the clothing worn, the temperature, the humidity level and the amount of air movement. With an air speed of 1–1.5 m/s, a 50% humidity level and an ambient air temperature of 25°C or 30°C, the cooling effect on the skin due to increased evaporation will be equal to a reduction of the ambient air temperature of 5.7°C at 25°C and 2.2°C at 30°C (from *Climate Responsive Building* by Paul

Gut and Dieter Ackerknecht). Evaporation can take place even though the relative humidity is 100% if the temperature of the air is lower than the skin temperature, but if the air has a higher temperature than the skin cooling effect by evaporation is not possible even though the relative humidity is less than 100% (from *Man, Climate and Architecture* by B. Givoni).

THE THERMAL ENVIRONMENT

Four factors make up the thermal environment. They affect the rate of heat loss and gain from the body, and together shape the thermal comfort zone. Here they are, however, described independently, and the thermal comfort zone is examined in a later section.

- Temperature: The air temperature range within which comfortable conditions may be established is approximately 16–30°C. Below 16°C excessive clothing or high activity rates are required. Above 30°C excessive air movement and sweating are normally required to maintain comfort even at low levels of activity. The temperature experienced by a person in a building also includes that from radiation from surrounding walls and possible direct radiation from openings. (Some books and studies use an operational temperature that is a combination of air temperature and radiation, and others use effective or perceived temperature, which is a combination of air temperature, radiation and humidity.)
- Humidity: Relative humidity of less than 20% is likely to cause discomfort because of the excessive dryness of the air. This may cause lips to crack, eyes to become irritated, and the throat to become sore. Relative humidity above 90% feels clammy and damp.
- Radiation: Heat gains due to direct radiation from the sun or indirect radiation from the surroundings will be the main source of discomfort in hot climates. Internally in a building, discomfort will occur when radiant temperatures, i.e. surface temperatures, are above the comfort range. This is because the difference between the surface temperature and skin temperature

will result in a heat gain being made by the body, taken from the warm building surfaces.

- Air movement: Wind speeds below 0.1 m/s may lead to feelings of stuffiness. Wind speeds of up to 2.0 m/s are acceptable when air movement is required, e.g. in a hot humid climate where other relief is not available. A wind speed of 1.0 m/s is normally considered as the maximum limit for night comfort. In an office, the wind speed and direction must be controlled to avoid high wind speeds where they create inconvenience, e.g. at working table level where papers can be blown about at a wind speed of approximately 1.0 m/s. The variation of the wind speed and direction (turbulence) can affect the comfort level. The cooling effect of air movements was mentioned at the end of the last section, Thermal exchange between the human body and its environment.

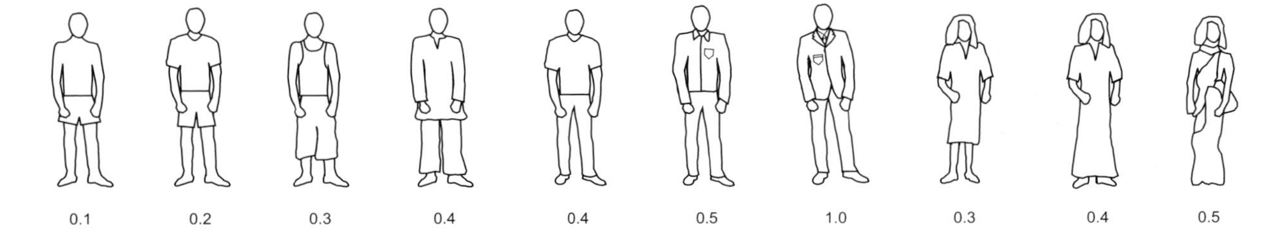

Thermal insulation properties (CLO units) of various items of clothing.

Clothing

Clothing offers humans thermal protection against environmental conditions in cold climates. In hot climates it protects against solar radiation, but it may also prevent the body from releasing excess body heat. Appropriate clothing in hot dry and warm humid environments depends on its design and material, i.e. whether it is light, loose and able to transport humidity. It is important for the cooling effect on the body that evaporation takes place on the surface of the skin and not on the surface of the cloth. The thermal insulation value, used internationally to measure the effects of clothing, is expressed in terms of CLO units (1 CLO = 0.155 m^2 °C/W). International standards have set one CLO equal to the value of a standard Western business suit. The CLO units for different clothing are given in the illustration above.

Generally, the amount of clothing that can be worn in hot climates to allow the cooling effects of air movement to be felt is, for Western outfits, equal to about 0.3 CLO units. However, for a loose traditional dress fashioned from a fabric of natural fibre with an open weave and in bright colours, CLO units of up to 0.5 will also allow the cooling effect of air movement. A long open dress may even create some chimney-stack effect, i.e. ventilation between the body and the dress due to air rising from the feet to the neckline (from *Housing Design in Extreme Hot Arid Zones* by Sami Elawa).

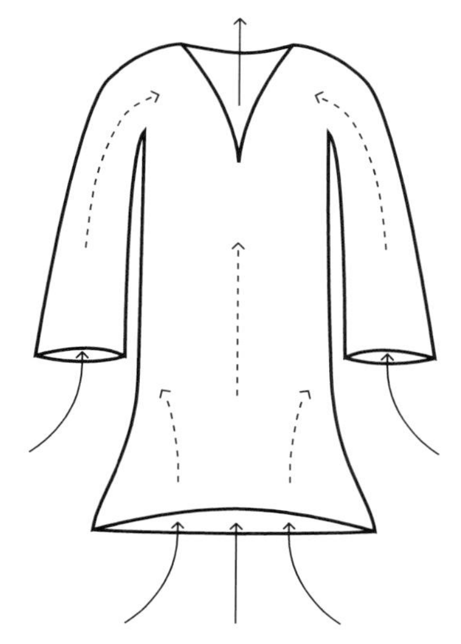

Evaporative cooling/ventilation in traditional clothing in hot environments.

Activity levels

Comfort conditions are related to activity levels and the resulting metabolic rate, i.e. heat production. Activity levels range from a minimum when one is asleep to a maximum when undertaking heavy physical work. The metabolic rate of the body is measured in MET, 1 MET being equal to 58 W for a sitting 'standard' adult in comfort conditions. For a person working in an office the metabolic rate is normally set to 125 W. Women generally have a slightly lower metabolic rate than men. The metabolic rate for physical work or sport activities may reach 1000 W or more. Certain activities can create conditions that fall well outside the comfort range. This may occur with strenuous activities of a short duration or with activities that occur at times of the day when temperatures are high. However, a degree of discomfort is acceptable under these circumstances. The body will be more susceptible to discomfort if these same levels are imposed upon it when resting, i.e. when it is due merely to high temperatures.

The metabolic rate is related not only to the level of activity, but also to an individual's conditions, such as blood pressure or stress.

Thermal comfort zone

The environmental conditions that are considered comfortable vary with the impact and ratio of the thermal environment factors (temperature, humidity, wind, etc.), as well as with clothing and level of activity as described above.

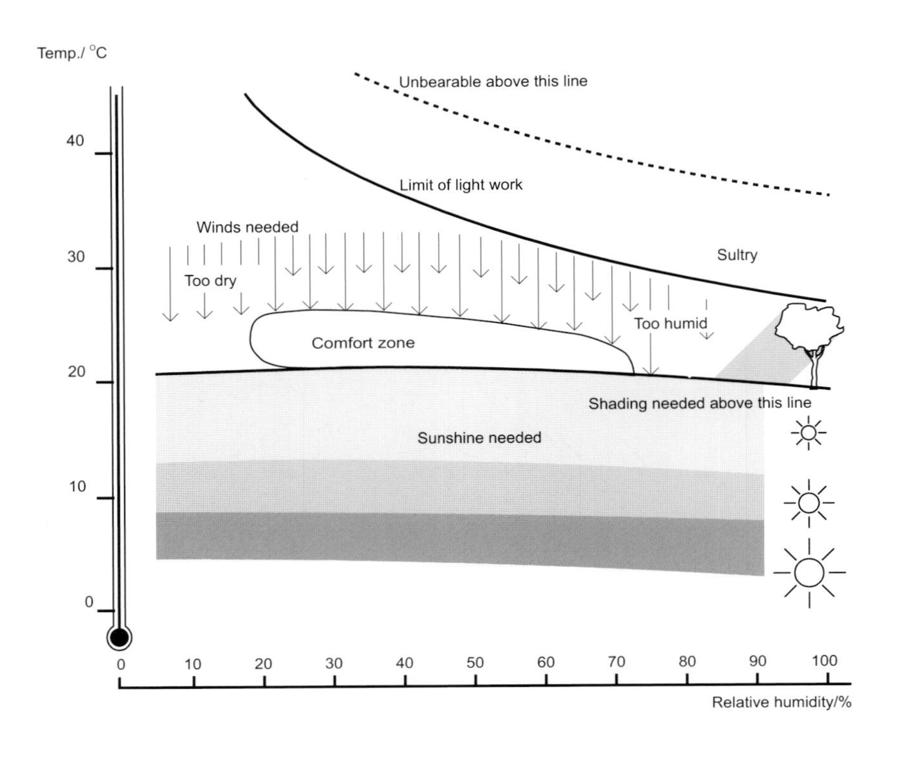

Thermal comfort zone by V. Olgyay.

The condition in which a person resting outside is supposed to feel comfortable, i.e. the comfort zone, is shown above. Victor Olgyay presented this diagram in 1963 in his book *Design With Climate*. His studies were carried out in a moderate American climate and based on temperature and relative humidity. Natural occurrences/passive measures, which can be exploited to achieve comfort externally in what would be uncomfortable prevailing conditions, such as sun radiation for low temperatures and cooling winds for higher temperatures, are also indicated.

In *Climate Considerations in Building and Urban Design*, B. Givoni suggested design strategies to achieve thermal comfort, shown on the next page. The diagram is based on thermal comfort inside a building, where not only air movement, but also the thermal properties of the building envelope and other passive cooling means, can be used to counteract high temperatures. The diagram also shows that, for combinations of very high humidity and temperatures found in warm humid regions, passive

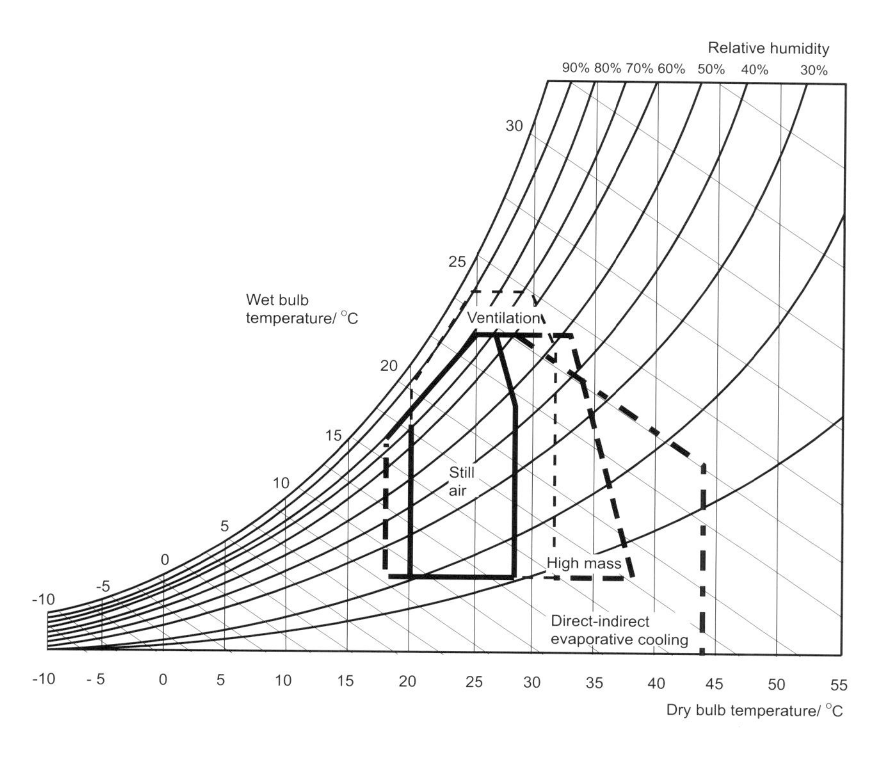

Diagram by B. Givoni showing strategies to achieve thermal comfort in buildings in hot climate. The diagram shows that in a dry zone with 20% relative humidity internal comfort can be achieved even with external temperature of more than 40°C when using high mass constuction and evaporative cooloing. In the warm humid zone with relative humidity of 70% external temperature of up to 30°C will be acceptable when only ventilation is used as a cooling agent.

means other than natural ventilation, thermal mass and indirect evaporative cooling will be required to arrive at an acceptable comfort level.

The comfort zone has been the subject of many studies. A common standard for air-conditioned offices today is temperatures of between 21°C and 26°C and relative humidity of between 30% and 70%, with air changes related to the use of each single room.

It has been found that occupants in air-conditioned environments are more sensitive to variations in temperature and humidity than people working and living in non-air-conditioned environments. In the latter type of environment, occupants are more tolerant of changes due to the natural relationship between changes in the internal climate and changes in the external climate, and they can control the internal climate to a certain extent, e.g. by opening a window, unlike the occupants of air-conditioned and sealed rooms. Situations with extreme climatic conditions that are outside the limits of the comfort zone can be found acceptable in non-air-conditioned rooms, depending on their duration and frequency.

Thermal comfort in non-air-conditioned environments varies mainly with the climatic zones. It is, for example, well known that people living in hot climates tolerate higher temperatures than people living in temperate climates, but they are sensitive to cool conditions.

For warm humid climates, field work by F. H. Mallick in Bangladesh (from *Architecture of the Extremes* by Y. Etzion (Ed.)) indicates a local comfort zone with a temperature range of 24–32°C, along with relative humidity values of between 50% and 90% with very little air movement. The activity level was between 0.8 and 1.2 MET, and the CLO values about 0.5. The study showed a particular adaptation to high temperature and relative humidity levels found in Bangladesh and sensitivity to low temperature and low relative humidity that is not found locally.

Thermal comfort studies based on laboratory research (from *Thermal Comfort* by P. O. Fanger) indicate that all humans react similarly to an artificial indoor climate or air-conditioning; the tests have been carried out in a laboratory on subjects living in different climates. However, for buildings regulated by passive measures and in real-life situations where adaptation influences the comfort zone, a locally perceived acceptable comfort must be used as the reference for thermal design. This will in turn reduce the possible needs for active measures, i.e. air conditioning.

Design principles: thermal comfort

- If it is too hot and dry there is a need for shade and cool surrounding surfaces for a person outside a building. However, if the air is humidified by simple means, the air temperature can be reduced significantly depending on the level of humidity and temperature. Air movement can reduce body temperature but, as mentioned previously, the effect is reduced with high temperatures. It can also be an irritant when the air is hot and not humidified, as it can increase heat gains and take moisture from the body at a higher rate than desirable. For a person inside a building the same applies, and the surrounding surfaces, i.e. ceiling, walls and floor, should not radiate heat. Entry of hot air from the exterior during the heat of midday should be avoided, but low temperature air from inlets with humidified or pre-cooled air will be useful. Measures can be introduced in order to reduce the surface temperature and radiation from the surrounding internal surfaces during the midday hours of maximum external temperature, e.g. by the shading of windows and external walls, or by the reduction and delay of heat flow. The following chapters give details of appropriate passive measures.

- If it is too hot and humid there is a need for shade and air movement to cool the body as well as the surrounding surfaces to which the body can lose heat by radiation. Air movements can reduce the body temperature by several degrees for most levels of temperature in this climatic zone. For a person outside or inside a building, cool surrounding surfaces are difficult to achieve during the daytime, but proper shading and other measures can assist in avoiding gains. For a person inside a building, the entry of hot air from the exterior during the very hot midday hours should be avoided, but the introduction of air with a low temperature or cooling by radiation, as well as simple fans to provide air movement, will be useful. The following chapters discuss the appropriate passive measures in detail.
- The aim should be to create internal climatic conditions that cater for typical stress conditions rather than to try to satisfy all or extreme combinations of thermal stress factors. It is important for thermal simulations to choose representative average meteorological data and appropriate comfort conditions. It is useful to keep in mind that the effects of air temperature and radiation from the surroundings plays an equal role in the experience of heat. If high radiation from walls is unavoidable the lowering of the air temperature is the only way to obtain comfort and vice versa.

Notes

When studying the comfort zone in a particular region or area it would be useful to examine where the population relax or work, both outdoors and indoors, during the various seasons of the year. This will indicate what qualities or elements of the climate the population appreciate. If office buildings introduce air conditioning, and if the employees live in non-air-conditioned private residences, the climatic changes during the day will be difficult to adapt to and may even present health risks for the employees. When air conditioning is introduced it will reduce the general comfort zone and separate the local population from their climate and culture, even though there are circumstances where heat and humidity make work, but not necessarily life, intolerable.

Also, it is not unknown that in the northern industrialised countries with cold climates, all climatic conditions cannot always be provided for, i.e. combinations of low temperature, snow and wind. Consequently, forced or uncontrolled climatic 'winter holidays' cannot be avoided in these regions, in spite of the fact that all sorts of sophisticated equipment for road clearing and an enormous amount of energy are used in buildings to avoid 'waste' of time or 'efficiency' on the part of society, owing to climatic conditions. One could expect a more appropriate attitude to extreme heat and humidity conditions in particular in the developing countries, where money and energy should not be wasted on cooling buildings in order to create working conditions in extreme situations. 'Heat holidays', changing working hours or fewer working hours due to not-viable-to-control conditions should be acceptable in the climatic zones discussed in this Guide.

3. The built environment

The built environment can be designed to take advantage of the beneficial aspects of the climate and modify those aspects that are unfavourable. It should focus on creating comfort not only for internal environments but also outside the buildings. Green areas in an urban setting and around the individual buildings will have an effect on the climatic conditions and will reduce the exterior temperature.

Below is a summary of the general principles that need to be considered in the design of the built environment. This is followed by design principles for the appropriate form and layout of the urban and the building environment as well as for the individual buildings. None of these principles can be adhered to blindly, and a combination or compromises will be needed at each specific site.

General considerations

Orientation

The orientation of urban forms and buildings must be based on their interaction with the sun and the prevailing winds. Since optimal solar and wind orientation seldom coincide, compromises will be required in the design process. It is the solar orientation of a building that determines the intensity of solar radiation that falls on the individual surfaces. In general, the following hold true.

- Horizontal surfaces will receive the greatest intensity of solar radiation.
- Of the vertical surfaces the east- and west-facing surfaces receive the greatest intensity of solar radiation. East-facing vertical surfaces will receive radiation in the morning, and west-facing vertical surfaces will receive it in the afternoon.
- North- and south-facing vertical surfaces will receive little radiation in latitudes close to the equator (and for short periods of the year).
- In the northern hemisphere, north-facing vertical surfaces will receive more radiation during the hot season.
- The south-facing vertical surfaces will receive less radiation during the hot season when temperatures are high and more in cool seasons when temperatures are low. South-facing surfaces receive more radiation in northerly latitudes. These conditions are reversed in the southern hemisphere.

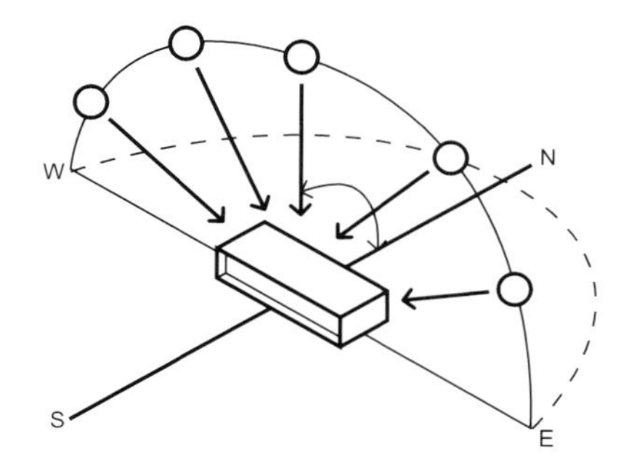

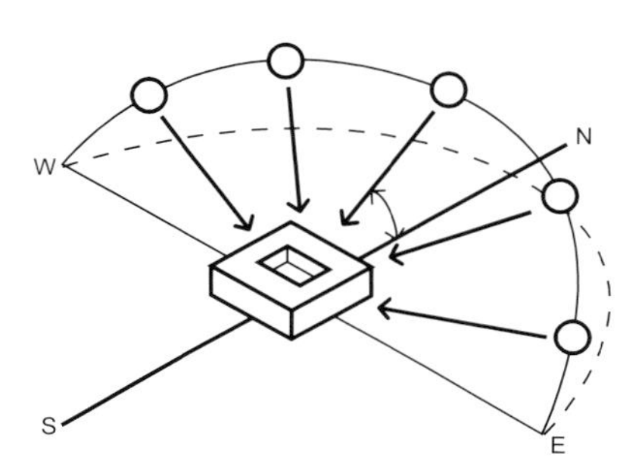

(top) At the equator the sun rises in the east, passes overhead (through or close to the zenith) and sets in the west. The east and west walls and the roof will receive most solar radiation; north- and south-facing walls can be increased in area without significant heat gains. (above) At latitudes north and south of the equator, the sun is correspondingly lower in the sky at midday. The south (or north) walls will receive more solar radiation.

Wind orientation will also need to be considered in order to maximise natural ventilation around and through buildings. The greatest pressure on the windward side of a building is gained when it is perpendicular to the direction of the wind. However, if openings are orientated at 45° to the prevailing wind direction, air velocities increase and the distribution of air improves within a building.

Conflicts between solar and wind orientation should be carefully analysed on each individual site. The geometry of the solar radiation cannot be altered, but the direction and velocity of air movement can be changed by the use of external elements *(left) Wind orientation to maximise air movement: a. airflow at 90°; b. airflow at 45°. In the second case the suction effect is increased, and internal air distribution is improved.*

(right) When there is a conflict between sun and wind orientation the wind can be redirected. a. Shows a building oriented east–west to minimize solar exposure, but the wind comes from the west, which will limit natural air movement in the building. b&c. Show possible solution by altering the wind pattern/air pressure around the building by projecting walls or staggering the rooms to facilitate cross-ventilation.

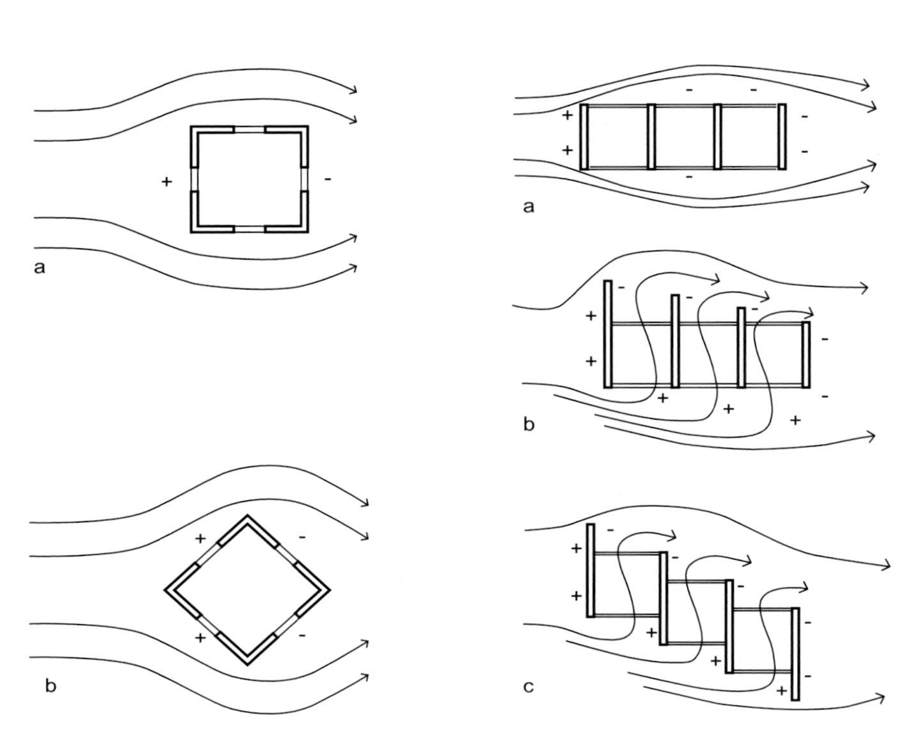

and building details, e.g. landscaping, or projection of roof and walls. (For more details see Chapter 7, Natural ventilation and cooling.)

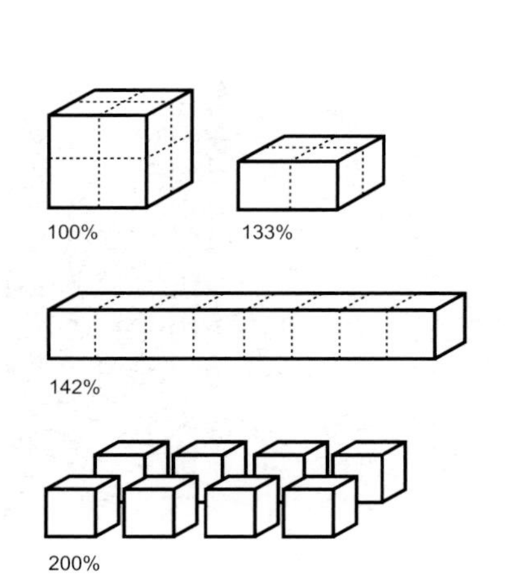

The surface area to volume ratio of various building layouts. The illustration shows surface to volume ratio when cubes are build as one cube and when spread out. The difference in percentage will be an indication of the difference in heat gains and loss due to surface.

Surface area to volume ratio

A building's ability to store or release heat is related to its volume (i.e. mass), whereas the rate at which it will gain or lose this heat is related to its surface area. The surface area to volume ratio will be an indication of the rate at which a building will heat up during the day and cool down at night. The height of a building, in particular in an urban setting, will also affect its heat gain and loss (see the section in Chapter 1, Building densities).

Spacing

The space between individual buildings as well as road widths will have an effect on the amount of solar radiation falling on or reflected onto buildings, on the air movement around and between buildings, and on the quantity and quality of light falling on the facades.

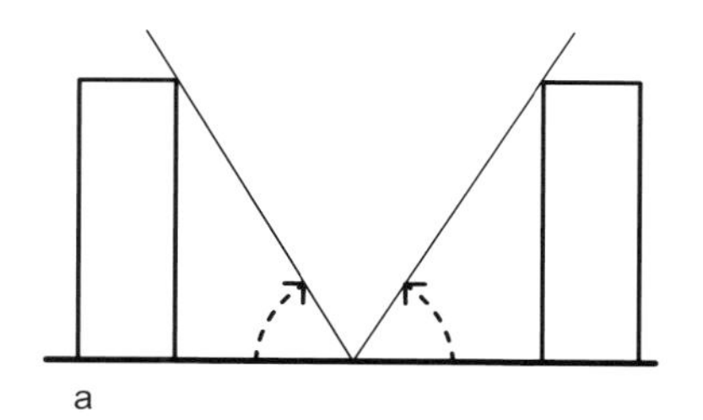

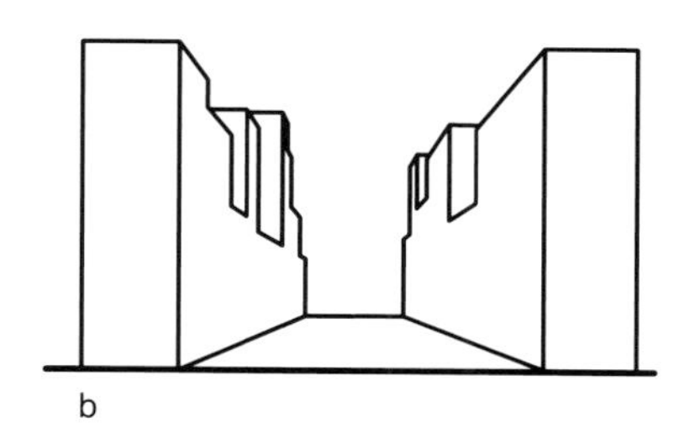

The effects of spacing between buildings: a. to control building heights, the depth of streets and external spaces;
b. to control sunlight and ventilation around sides of buildings rather than over roofs.

Design principles: the urban environment

Hot dry zones

The overall urban design objectives in hot dry zones should aim at:

- providing maximum shade and minimising heat gains (proper solar orientation is more important than wind orientation);
- minimising reflection and glare from indirect solar radiation in streets and open spaces, and moderating the effects of undesirable winds.

The main urban design principles are as follows.

- Compactness in form and layout to minimise sun exposure will allow buildings to shade one another and create shaded and outdoor spaces. The buildings should be uniform in height in order to avoid uneven distribution of wind. Heat island effects should be avoided (see the section in Chapter 1, Building densities).
- Narrow winding street networks should be used. This layout will minimise solar radiation exposure, reduce the effects of stormy and dusty winds, and create shade areas throughout the day.
- The incorporation of street planting, green gardens or patios, and pedestrian networks protected by facade treatment (e.g. colonnades, awnings) are specific protection constructions.
- The use of the courtyard form and layout (see Special topic: courtyard design at the end of this chapter) on the urban scale will maximise both shade and ventilation.

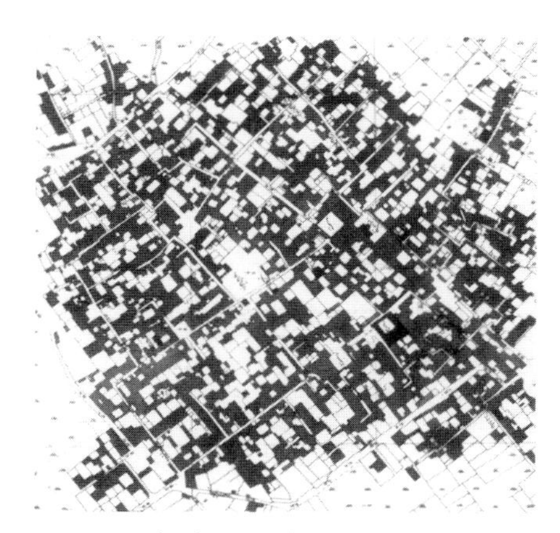

An aerial photo of an unknown city in Omar. This compact urban layout is found in hot dry areas. Its narrow winding streets reduce the effects of stormy winds and its public and private courtyards reduce the area's overall surfaces exposed to solar radiation.

The urban form should be compact to expose little to the sun except the roofs of buildings.
(top) Photo from Tunis, Tunisia.

(below) photo from Yazd, Iran.

Narrow streets will give shade for pedestrians and on buildings.
(top left) Cairo, Egypt. (top right) Apt in the South of France.

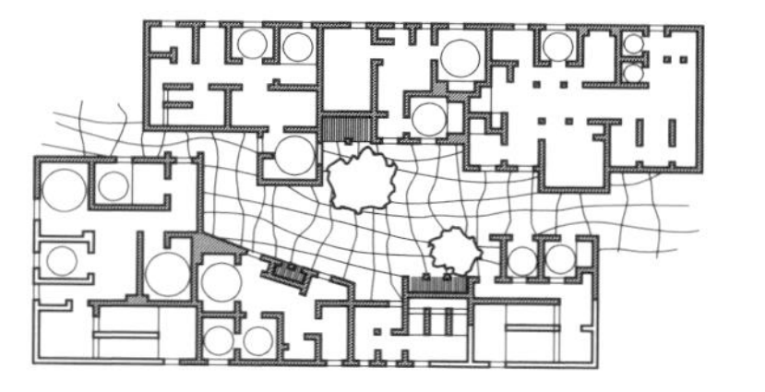

(left) Drawing of a pedestrian street in Harraniya Village, Egypt, showing how interconnected an urban fabric can be, woven with individual courtyards and still maintaining privacy. Based on scheme design by Architect Hassan Fathy.

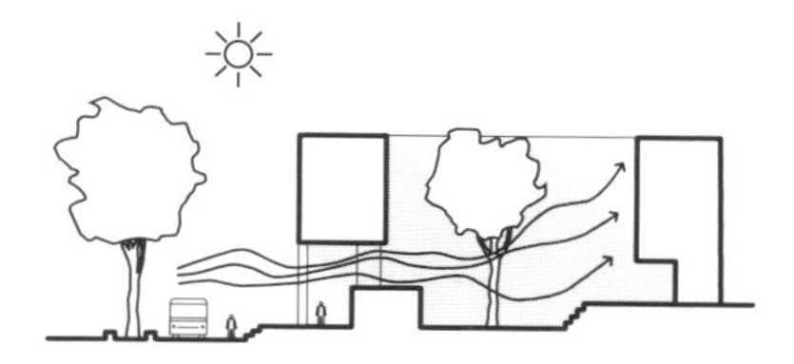

(bottom left) A drawing illustrating means of creating cooling cross-breezes at pedestrian level and maintaining a compact urban layout.

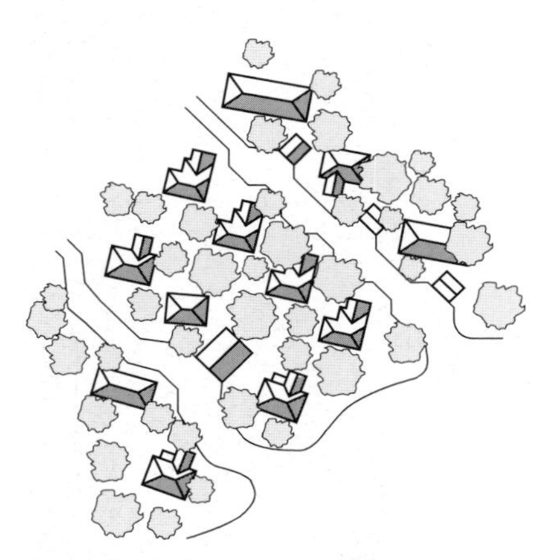

An urban layout suitable for warm humid regions. The buildings should be oriented to maximise the benefits of the prevailing cooling winds.

(below) A settlement of houses on Bali, Indonesia. The layout and form is planned to catch the prevailing winds. The houses have an open structure to allow cross-ventilation, and the roofs protect from radiation and heavy rains.

Warm humid zones

The overall urban design objectives in warm humid zones should aim at:

- providing maximum shade and minimising heat gains; and
- maximising the potential of any air movement to create natural ventilation. Wind orientation is more important than solar orientation.

The main urban design principles are as follows.

- The urban form and layout should promote openness. Buildings should be widely spaced to encourage air movement between and through all buildings.
- Buildings should be orientated approximately perpendicular to prevailing wind directions.
- Streets should be relatively straight and open to encourage air movement. They should be lined with tall, high-canopy trees to provide shade and not restrict air movement.
- Landscaping and architectural elements should be carefully designed to create comfortable conditions at pedestrian level.

Examples from warm humid zones. (top left) Public buildings in Queensland, Australia, by architect Bligh Voller. Roof overhangs and balconies are used to shade the buildings and the external spaces. (top right) Open and light buildings in an urban housing environment with a composition of shade-creating features: roof overhang, deep balconies and awnings. From Rainbow Shores Estate, Queensland, Australia. Architects: Glare design. (below left) Pedestrian-friendly streetscape from Durban, South Africa. It provides shade and is open to air movement.

Notes: Low-cost urban planning and housing schemes

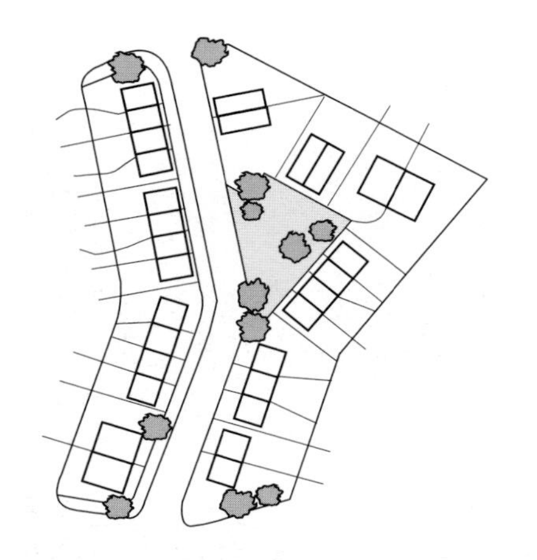

Cluster of houses

Illustrations from a low-cost housing project in Port Elisabeth in South Africa.

In most countries of the world urban planning exercises are limited to the planning of buildings in existing urban environments or to planning isolated business and/or shopping centres outside the existing urban structure. Many new buildings, even those in an existing urban environment, are designed with the main objective of giving an individualistic appearance and not as part of the microclimate around them. They may have some passive means of ventilation and cooling, depending on the interests of the owner and rules and regulations. However, many low-cost housing schemes are planned in the countries within the climatic zones of the present Guide, but in most cases without regard to the fundamental principles of climatically adopted design, e.g. by introducing small and uniform lots on parallel rows. Each site is fenced or walled off, and a small house is placed in the middle of the site. The layout impedes the use of active design principles, e.g. compactness for hot dry regions, and openness for warm humid regions. This may turn the inhabitants into future consumers of energy for air-conditioning equipment in order to counteract the uncomfortable internal climate in the houses. Planning is often done together with the inhabitants, a process that is used to legitimise the work of the developer, but the inhabitants are unfortunately not aware of the consequences of this type of planning and design of the buildings.

An example of interesting planning of a low-cost housing scheme constructed in Port Elisabeth in South Africa is shown. Here semi-detached houses of up to two storeys have been introduced. It was found that the plan allowed each family to have up to 56 m^2 interior space compared with 30 m^2 in an individual house for the same construction budget. The site plan allows for active design principles and has also introduced opportunities for an enhanced social environment compared with fenced houses 'in a row'.

Design principles: the building environment

Hot dry zones

The main principles are as follows.

- Buildings should be compact and inward-facing to reduce surface areas exposed to solar radiation.
- Large surfaces should be orientated north–south, as these will receive the lowest solar radiation exposure. West-facing surfaces are the most critical as peak solar radiation intensity coincides with the highest temperatures in the afternoon.
- Individual buildings should be grouped closely together for mutual shading of exposed surfaces.
- Access to cooling and dust-free winds should be promoted.
- Small, enclosed courtyards can be designed to create usable, protected outdoor spaces.
- The shading of buildings and outdoor spaces is critical. Projecting roofs, verandas, shading devices, trees, surrounding walls and buildings are just some of the ways to achieve this.

(top) Buildings and outdoor spaces should be introverted and provide shade. Photo from a building complex by architect Jan Utzon in Shamwa, Zimbabwe.

(above) The buildings can have shading provisions such as planting and colonnades, as shown on this photo from Ouagadougou, Burkina Faso.

(left) Building forms should be compact and have small and well protected openings towards the exterior as illustrated on this photo from Brisbane, Australia.

Examples of buildings and outdoor spaces that are introverted and provide shade. Photos from (left) Yazd, Iran, and (right) Tunis, Tunisia

Warm humid zones

The main principles are as follows.

- Buildings should be open, outward orientated and elongated to take advantage of all air movement. Surface areas should be large compared with the volume.
- In terms of solar orientation, buildings should be elongated along an east–west axis.
- Wide spacing between buildings allows for ventilation of external and internal spaces.
- Openness and shading should be the dominant features of buildings. All vertical surfaces and openings should be protected without restricting air movement.
- The benefit from the wind (the air speed) will, in general, increase with the height of the building.
- Landscaping and architectural elements should be incorporated to create comfortable indoor and outdoor environments. Projecting roofs, verandas, shading devices, planting, etc. can be designed to provide shade without restricting air movement.

A radical and simple example of climatic response to hot humid climate is this chapel building on Komalie Farm, NT, Australia, placed in a shaded but open environment. The building itself provides the main characteristics needed, i.e. shade and cross-ventilation. However, the corrugated steel elements are not the best material owing to its heat-absorbing property, but the elements are fortunately not much exposed to the direct sun owing to shade from the trees. The inclination of the elements will enhance the cooling effect from the ventilation taking the cool air from the ground and directing it upwards to the occupants.

(right) A typical traditional Japanese house from Kyoto, Japan. It is surrounded by trees, and has openings that can serve to control ventilation and light. It has a light wood structure and is elevated from the ground.

(middle & bottom right) These two examples are from Darwin, Australia, by Troppo Architects. The buildings are spaced apart from other buildings to encourage air movement. They are narrow to allow cross-ventilation, have large roofs and awnings for shading, and are lifted off the ground. They are constructed in wood to avoid heat storage.

Special topic: courtyard design

The courtyard model can be defined as a building, a group of buildings or building elements surrounding an internal space that is open to the sky. The typical North African/Middle Eastern courthouse building is such a model. Its response to the demands of a hot dry climate is very appropriate and is evidenced by examples using the same principles in most hot dry regions around the world. Other and more open and larger courtyard buildings, e.g. the haciendas in Mexico and Spain or the Chinese and Japanese models, respond to different climates and cultural issues of privacy as well as to different functions of the internal court. The North African/Middle Eastern courthouse building represents a dual response to the problems of solar radiation and a windy, dusty environment. It is also an excellent thermal regulator. Its high walls cut off the sun and shade large areas of the inner surfaces and the courtyard floor during the day. The cool air, the surfaces and the courtyard floor draw heat from the surrounding areas and re-emit it to the open sky during the night.

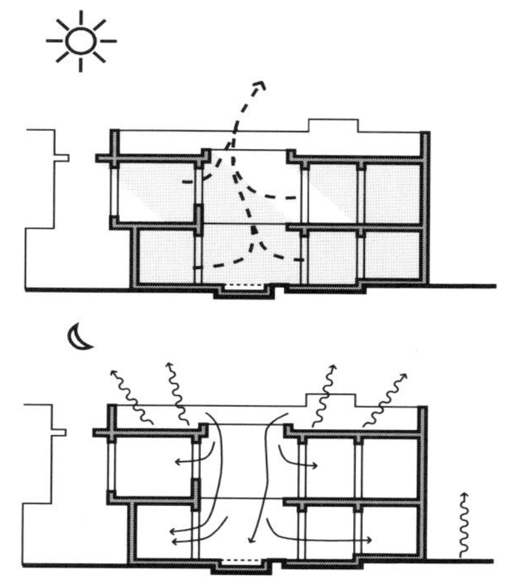

The courtyard model's response during the day and night.

The outdoor spaces created by the courtyard model should be designed to achieve the desired amount of daylight penetration, reduce solar heat gains and promote cooling breezes, but exclude hot, dusty winds. They should also be incorporated as part of the overall living spaces of a city or an individual building.

The courtyard model can be used at both the urban scale and for individual buildings. It has proved particularly effective against solar penetration when buildings are grouped together to create an entire urban fabric, as fewer surfaces will be exposed to the sun.

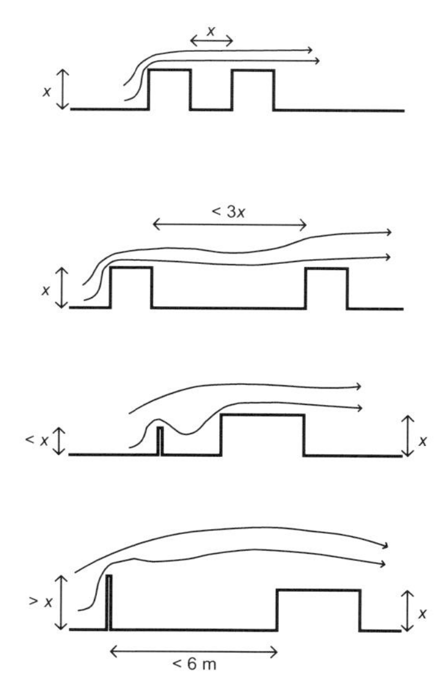

The design of various courtyard spaces. A square courtyard offers good protection from wind-blown dust and sand. The depth of a rectangular courtyard must be restricted (3x) unless the long axis is perpendicular to the wind. Barrier screens must be of a suitable height and not more than 6 m from the building to provide protection. From various sources.

Its general form should be one that turns its back on the extremes of the surrounding environment and brings its focus into the inner realm. All openings should face onto the courtyard space. Its outer walls should be blank with few or no openings (i.e. only those required for ventilation).

The defined, enclosed outdoor space of the courtyard creates a place of sanctuary and privacy. It also has the distinct advantage of producing a pleasantly cool external environment and one that

(above) The courtyard offers sanctuary and relief from a hostile external environment. The landscape treatment of these spaces can greatly increase the cooling potential of the air. A photo from Cairo, Egypt.

(right above) Courtyard buildings from Shamwa, Zimbabwe, designed by architect Jan Utzon.

(right below) From Granada, Spain. Both use colonnades to provide protection from the heat of the day as well as fountains, pools, and vegetation to cool the air.

assists in the cooling of the internal environment.

During the day the courtyard space heats up quickly compared with the buildings around it. This will create a stack effect, i.e. the movement of air due to height and temperature differences. As the hot air rises, it will draw air into the internal spaces, thereby setting up a breeze. If this air passes over water or planting on its way into the building, the cooling effect will be even greater.

During the night the courtyard space acts as a sink, collecting the cool air flowing down from the roof. This cool air will filter in through doors and openings facing the courtyard.

The courtyard building will always have one of its external walls in shade. Care has to be taken to shade the inward-facing walls of

Courtyard buildings and air movements in a protecting (oasis) environment. Based on illustration from Back to Earth: Adobe Buildings in Saudi Arabia *by William Facey.*

the building from direct exposure to the sun. This can be achieved by using colonnades or verandas. These intermediary spaces will act as thermal barriers.

The courtyard space will allow light to penetrate a building if there are no external windows. If a colonnade surrounds the courtyard wholly or partly, light will be filtered and relieved of glare. Planting will improve this further by reducing the glare created by the bare ground.

Another virtue of the courtyard model is that cooling breezes are normally available regardless of wind directions. Winds will pass over the building and create areas of low pressure in the buildings and the courtyard. The air flow through ventilation openings (i.e. on the outside walls) will move to these areas of low pressure, thereby creating a flow out through the rooms and into the courtyard. This movement is further reinforced by the stack effect.

Plants and trees outside the building play a vital role when the wind is forced to pass through them, thus allowing winds to be cooled and relieved of much of their sand and dust.

4. The external environment

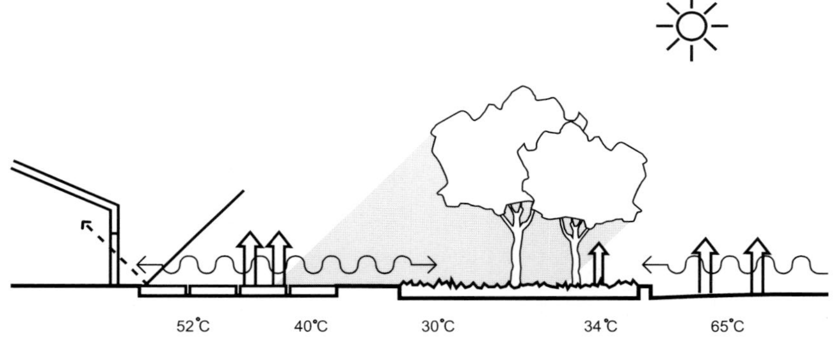

The temperature in and around buildings will be affected by the nature of the surrounding surfaces. The temperatures shown were recorded in a hot dry climate when the air temperature was about 40°C.

Active design does not end at the building. It also extends to the surrounding spaces, where landscaping will affect the climate internally in the buildings. Trees, bushes and grass have a cooling effect due to evapotranspiration from their surfaces which cool the ambient air, a shading effect and a low reflectivity of heat. Pavements, which absorb and reflect heat, must be avoided. The external environment is particularly important in the hot climatic zones, where most people spend a great deal of their time outdoors. The outdoor areas should be treated as an extension of the living and working spaces.

Enclosure walls provide privacy and protection between the garden or the courtyard and a hostile external climate. Photo from Honolulu.

Design principles: the outdoor spaces

Hot dry zones

In hot dry climates, adjacent buildings, pavements and bare ground heat up quickly, which both creates painful glare and reflects heat towards buildings. Protection against hot and dusty winds may also be needed. The following principles can be adopted to improve these conditions.

- Shade: Daytime use of outdoor spaces is possible only when they are shaded. Walled gardens, enclosed courtyards, loggias, pergolas and landscaping can be used to cut out direct sunlight and glare from reflecting surfaces.
- Enclosure: Enclosures can protect against hot dusty, winds. They also provide privacy and a physical demarcation between the garden or the external private space and the hostile external climatic conditions. Enclosures can be created by using structures and planting.
- Courtyard planning: This will reduce heat gains in the spaces and buildings that surround them. If the floor of the courtyard is shaded for most of the day and open to the sky at night, it will lose more heat by outgoing radiation at night than it will

Vegetation to improve the micro-climate and reduce glare in a court-yard on Honolulu.

gain during the day. It will also be the coolest part of a building and will absorb the heat from the adjoining rooms. A shaded courtyard with landscaping creates a comfortable extension of the indoor working or living space.

- Evaporative cooling: As scarce as it may be, water is always an asset in dry climates. The physical effect of evaporation is essential, but the psychological effect of water is even greater. The sound of water is known to bring relief long before a measurable reduction in air temperature can be recorded.
- Plant selection: This should be restricted to those species that can survive with very little water and to areas where they can be maintained. As bright reflection comes from bare ground, it is essential to have green areas to relax the eye. Ideal plant species include evergreen trees, shrubs, various ground-cover plants, perennials and some special grasses. Climbing plants can be used to cover walls or fences or for other specific purposes trained on wires.
- Protected circulation: Preferably, no one should be forced to walk in the sun. Pedestrians need shade and protection. Narrow streets, arcades and covered pavements are preferable to wide,

(left) An operable lightweight canvas awning or similar is useful for protection during the day and when removed during night for cooling of external spaces, as in this picture from a restaurant in San Sebastian, Spain.

(top & above) Colonnades and protected walkways can be used to create shaded outdoor spaces as well as reduce heat gains and glare. Photos from Alger, Algeria.

open roads. Cars need to be provided with covered or shaded spaces.

- Public outdoor spaces: Generally, the principles used to create outdoor spaces are the same for both private and public areas. Public spaces must also be enclosed, inward looking, carefully landscaped, cooled by water and shady. Their size must also be related to the available facilities for watering and maintenance. If they are too large, the green recreation areas, the lungs of the city, become barren and unusable patches of desert.

Warm humid zones

The inhabitants of warm humid regions spend a great deal of their time outdoors. Outdoor life can be pleasant if provisions are made to facilitate air movement and shade and provide protection from the rain. These provisions will also improve internal comfort.

The design principles are as follows.

- No enclosure walls: If property needs to be demarcated, fencing or perforated screens that obstruct views but not air movement should be used.

External shading from the roof and the building structure at the University in Brasilia, Brazil.

- Open planning: The need for breeze rules out courtyard planning or long lines of buildings flanking a street. Buildings should be grouped freely and as independently as possible. At high urban densities increased height is preferable to increases in ground cover. Buildings may be used to draw breezes into an area, but may possibly at the same time prevent air movements in spaces behind them. It is more appropriate to raise rows of buildings on stilts and in general to make them as 'wind transparent' as possible.
- Shade: People who work, play or rest in the open require sun protection. Roof overhangs, verandas, porticoes, covered walkways and landscaping are welcomed. The open spaces left under elevated buildings can be used as shaded outdoor spaces.
- Planting: Shade-producing trees should have a high priority in the design of open-air living spaces. They filter the sunlight,

reduce air temperature through evaporation, and reduce glare from bright overcast skies. The varieties of trees that thrive in warm humid climates offer a range of opportunities. Equally important is ground cover but, if it is abundant, care has to be taken to remove or control it in order to keep areas vermin free and to maintain air movement around buildings.

- Shelter from rain: Since rain showers are frequent in warm humid zones, covered verandas where work and play can continue during these periods should be planned. Verandas or roof overhangs provide rain protection for open-wall buildings used in the warm humid zone. Shelter from rain includes covered pedestrian passageways along streets and between buildings. Trees and overhead climbers can also be used to provide protection.
- Public open spaces: Unkept open spaces in warm humid zones are usually green. However, it is more effective not to allow any incidental or undefined open spaces to emerge. Open spaces should be designed to serve a particular purpose in order to prevent their misuse. It is therefore important to designate and demarcate open spaces clearly and to allocate responsibility for their maintenance.

(top left) Shading trees for a walking street in Sydney, Australia.

(top right) Simple shading structure with bamboo matting from Lagos, Nigeria.

(above) An open shaded market in Basse-Terre, Guadeloupe.

5. The building envelope and its components

It is at the building envelope that the interrelationship between the given external conditions and the required internal conditions are determined. The building's envelope must respond to a wide and varying range of external conditions. The most important element in hot climates is solar radiation heat gain on the building envelope and the subsequent release of heat by radiation from the inside surfaces towards the occupants of the building.

The building's envelope needs also to be able to release internal heat gains. These may include heat from warm internal surfaces, humans and activities, appliances, electrical equipment and artificial light fixtures.

Compromises will be necessary to achieve the appropriate thermal requirements, ventilation requirements and daylighting levels in order to satisfy a building's functional requirements.

THE OVERALL BUILDING ENVELOPE'S PERFORMANCE

Hot dry zones

The main function of the building envelope in hot dry zones is to reduce heat gains and address the large variations between day and night air temperatures. It will require thermal mass to store heat during the day. Heavyweight or composite construction can be used to create this mass. However, the heat stored during the day will need to be removed at night. This cooling down process cannot take place through heat loss (i.e. long-wave radiation release) by the building envelope alone. It has to be assisted by the night-time ventilation of internal spaces.

Warm humid zones

In warm humid zones air temperatures vary little between the day and night. Therefore the building envelope cannot cool down sufficiently at night to allow the storage of heat during the day. It should be of a lightweight construction to reduce the storage

Special envelope performance. The Botswana Technology Centre, Gaborone. Hot dry climates. The pictures show the central circulation space. The envelope is designed around this central space with highly placed louvred openings to the exterior creating an internal street with natural ventilation and indirect light. The internal street has fountains with falling water, which moisten and cool the air. The pre-cooled air is used in an advanced evaporative cooling system, which cools the whole building. Architects: Anderson & Anderson International in conjunction with architect Stauch Voster. Mechanical and Services Engineers: Arup, Botswana and Ove Arup & Partners, London. Structural Engineers: Anthony Davenport & Associates.

Special envelope performance. A professional training centre in Herne-Sodingen, Germany. This project illustrates the possibility of an outer skin as the first line of defence against the external climate, thus limiting and controlling the impact on the internal envelope. The glassed external structure has among other direct solar radiation protection and openings to control ventilation. The roof has large solar energy collectors, which at the same time provide solar protection. Architects: Jourda & Peraudin, Lyon, France. Engineers: Ove Arup & Partners, London. For hot climates discussed in this Guide similar outer-skin envelopes may be feasible, in particular with the invention of transparent insulation materials.

of heat by the structure. It should also be as transparent and open as possible, to maximise the cooling effects of air movement throughout all internal spaces.

THE IMPACT OF SOLAR RADIATION ON THE BUILDING ENVELOPE

There are three ways in which a building experiences heat gains and losses due to solar radiation:

- radiation enters directly through openings and is absorbed by its internal surfaces, thereby creating a direct heat gain effect;
- radiation is absorbed by the external building surfaces (i.e. opaque) and transferred to internal surfaces; this involves a heat exchange process, which also results in heat gains; and
- radiation gained by a building may also be released back into the atmosphere. This occurs when the sky temperature is cooler than those of the building's surfaces. This may occur during the day and night, though more readily at night when temperatures are lower and the sky is clear. This heat transfer process will result in a building experiencing heat losses (long-wave radiation).

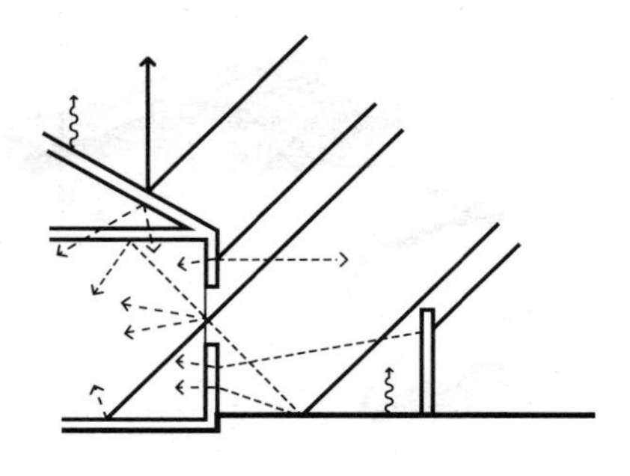

Heat exchange due to radiation between the building and its surrounding environment.

Table of solar radiation impact (%)

Condition	Direct: on a horizontal surface	Diffuse: from the sky (solar angle)			Reflected: from the ground on a vertical surface
		Low	45°	High	
Sea level, clear sky	100	20	11	7	10
1000 m altitude	110 (108–114)	18	10	6	11
2000 m altitude	122 (117–128)	17	9.5	5	12
3000 m altitude	127 (123–130)	16	9	5	13
High humidity, dusty or polluted	78	Average 12			8
Hazy sun and thin cloud	43	Average 23			4
Overcast	20	Average 25			2
Heavily overcast	5	Average 30			1

(From *Housing Climate and Comfort* by Martin Evans.)

The intensity of solar radiation varies according to the orientation of surface, the altitude of the sun, the latitude and the climatic variations.

It can be observed from the table that solar radiation is more diffuse than direct in warm humid regions where the sky is often overcast. For a town such as Durban in South Africa, diffuse solar radiation represents, on average, 45% of the total solar radiation over the year (from *Manual for Energy Conscious Design* by Dieter Holm). In such conditions solar protection becomes more complicated, e.g. window or roof overhangs designed to protect the opening or the wall from direct radiation will not be adequate to provide protection from diffuse radiation that comes from all directions, i.e. the whole sky. The solar radiation impact value is fortunately lower when the radiation is diffuse than with direct radiation.

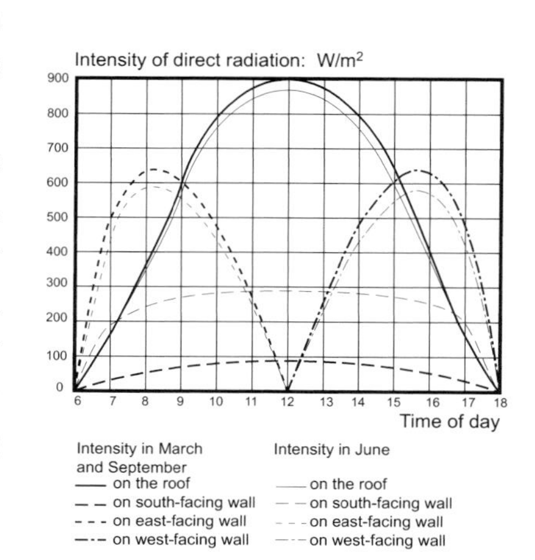

An example of solar radiation intensities for various surfaces at latitude 5° north.

At night the sky radiates long-wave radiation to the earth but at a rate less than the long-wave radiation from the earth to the sky. The effect of this heat exchange is significant in hot dry climates and lower in warm, humid climates owing to cloud cover, which restricts radiation from the earth to the sky.

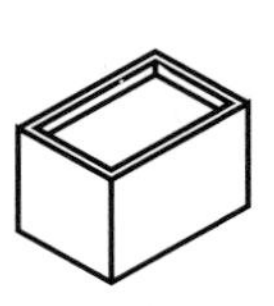
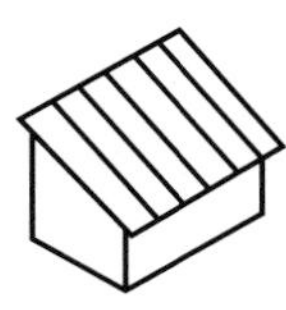
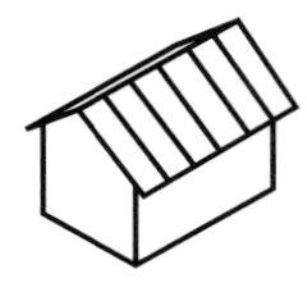

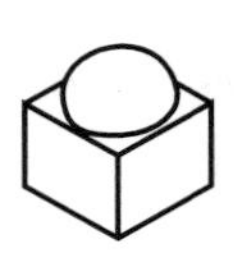

Roof design

The most important building envelope component is the roof as it is exposed to the greatest amount of solar radiation and it is the most difficult to protect. Its performance will depend on its form, construction and materials. A roof will respond to climatic conditions in the following manner.

When outdoor air temperatures are higher than indoor air temperatures, the outer surface of the roof absorbs radiation and heats up. It transfers this heat to its inner surface, where it raises the temperature of the air with which it is in contact (i.e. through conduction). This heat in turn radiates out and is absorbed by people and objects within a building.

In composite roofs with a separate roof and ceiling, the heat transfer between the two surfaces is partly radiated and partly conducted. If the space created between the roof and ceiling is enclosed, the trapped air may reach a very high temperature, which further increases the transfer of heat.

Different roof forms will have varying effects in terms of dealing with specific climatic conditions. The selection of the appropriate roof form will have to address the need for:

- protection against solar radiation exposure, to reduce heat gains;
- air movement across its surfaces, to aid in the cooling process; and
- rain protection to shed water readily from its surface and away from walls and openings.

Roof design and heat gains

Flat roofs experience exposure to solar radiation throughout the day. Consequently heat can be transferred into the building from the whole surface throughout the day. These heat gains will be excessive if measures such as a ceiling and air cavity or roofs of heavyweight construction (i.e. to store heat) are not incorporated.

During the night the temperature just above a flat roof is lower than that of the surrounding air temperature owing to emission to the sky. This can be used to cool internal and external spaces.

Cool air will remain at the roof's surface because of its increased density, but it can flow down into spaces that are warmer. A low parapet wall can be used to prevent the cool air flowing off the edge of the roof to areas where its benefits will be lost. A simple method of directing the cool air is to use a roof that is slightly sloped towards an enclosed courtyard. The courtyard will act as a basin, trapping the cool air and allowing it to flow into the rooms that surround it.

An advantage that flat roofs present is that they can be designed to be functional spaces (e.g. a terrace or outdoor sleeping area).

Arched, domed and pitched roofs moderate heat gains as less of the roof's surface is exposed to the sun. During the day, rounded forms, unlike flat roofs, always have part of their surface in shade. The part that is exposed to the sun heats up and transfers this heat to its inner surface. This inner surface readily loses this heat to the cool surface, which is in shade. This heat exchange process also occurs in pitched roofs when part of the surface remains in shade during the day.

The large surface area of rounded and pitched roofs also provides a large area over which convection, heat exchange and long-wave radiation heat loss can occur, which will reduce heat gains during the day and increase heat release during the night.

These roof forms can also be designed to increase the height of the rooms below, which can be used to create spaces far above the heads of the occupants into which warm air can rise.

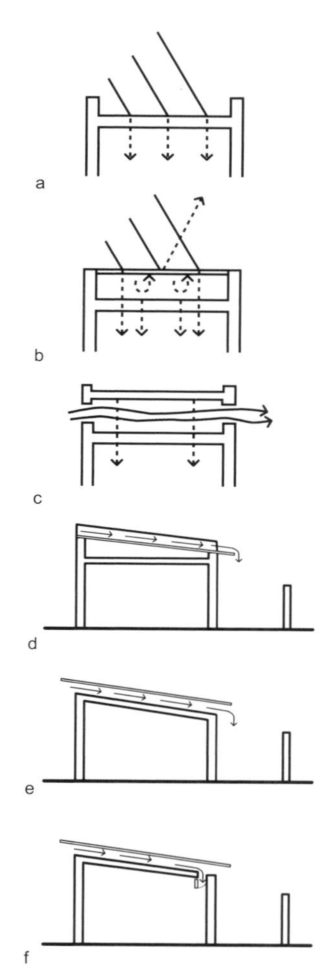

a. Flat roofs will be exposed to solar radiation throughout the day. b. If a reflective outer roof and an air cavity is incorporated, solar radiation will be partially reflected and partially trapped in the roof space. c. Roof ventilation will remove trapped heat. d. A sloped roof surrounded by a parapet wall can be used to draw cool air down into a courtyard space at night. e & f. The sloped roof can be separated to allow free air movement under the roof, which will increase the cooling process further. f. The cooled air can be directed to the interior of the building.

(top right) A domed roof will always have part of its surface in shade during the day. At night a domed roof's large surface area means a greater area from which long-wave radiation heat loss can occur. A similar phenomenon applies for a steep pitched roof.

(below right) Examples of air movement across various roof forms. They show how sharp edges make the air stream depart from the roof (and cause a suction effect), unlike a round form.

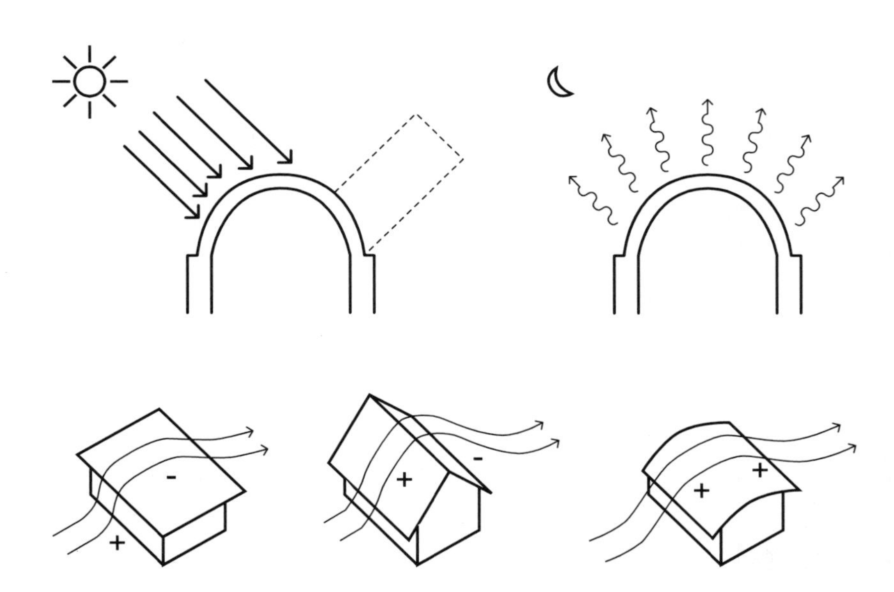

Roof design and ventilation

The transfer of heat through the roof can be reduced by allowing air to move readily across its exposed surfaces and by ventilating the roof space. The first method involves the appropriate orientation of the roof, whereas the second involves the use of composite roofs with ventilation openings.

Roof slopes should generally be orientated towards the prevailing breezes, and obstructions to the air movement should be avoided.

The provision of ventilation in composite roofs serves two functions: the removal of heat, and control of the moisture and humidity in the roof space. Ventilation of the space between the roof and the ceiling can be achieved through the provision of ventilation openings. Care has to be taken that they are properly designed, as the size and location of the openings will affect the quantity and speed of air moved through the roof space.

(above left) A roof's orientation and form can be designed to catch the wind and to cool external surfaces. The building is from Darwin, Australia.

(above) Roof ventilation systems on the French Cultural Centre in Maputo, Mozambique.

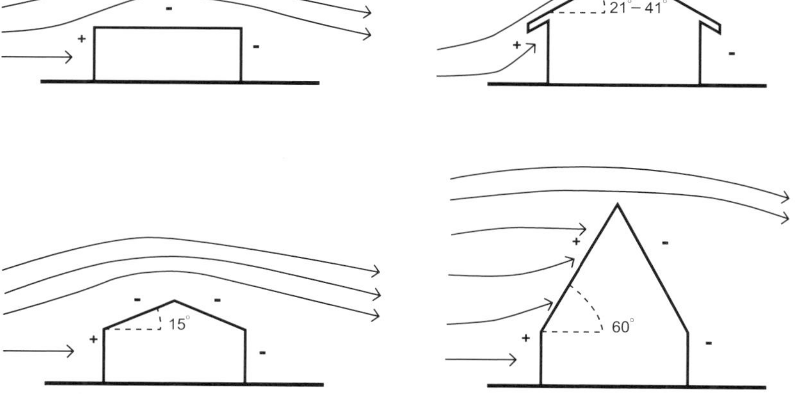

There is negative wind pressure (suction) on the windward side of a double-pitched roof up to a roof angle of 21° regardless of the wind direction. Between 21° and 41° there is a variation of low positive and negative pressure depending of the wind direction. Only at 60° will there always be pressure on the windward side. The above information is important for placing ventilation devices in the roof. From Natural Ventilation in Buildings, *Francis Allard (Ed.).*

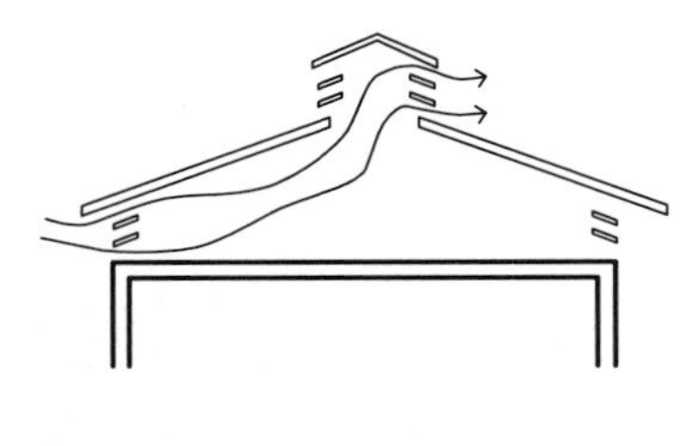

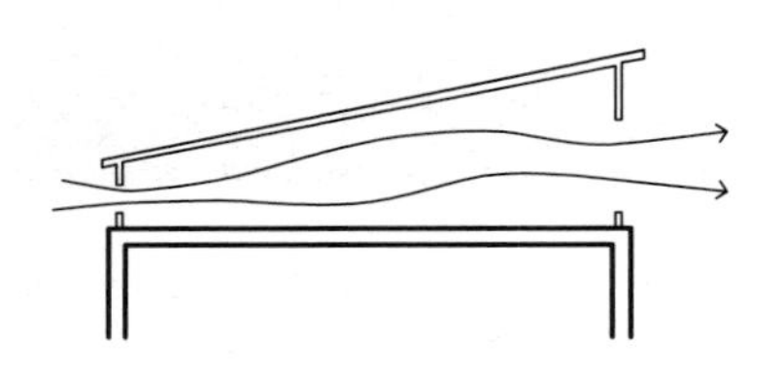

(top) Height differences between openings in the roof space will allow warmer air to rise and escape through the higher openings and be replaced by cooler incoming air.

(above) The air outlet opening should generally be larger than the air inlet opening, to create pressure differences that will draw a greater amount of air out of the roof space.

The air outlet opening should be larger than the air inlet opening in order to create a maximum pressure difference so that the greatest amount of air is drawn out of the roof space. This rule should be applied to the prevailing direction of wind during the hottest season. However, it is normally advisable to use relatively large openings for inlet and outlet owing to variations in wind direction.

Ventilation openings should be located to create a difference in height between inlet and outlet openings. This will allow warm air to be drawn out of the roof spaces simply as a result of the temperature difference. Lightweight roof construction such as corrugated metal or fibre cement sheeting may suffer from internal condensation problems. Efficient ventilation of the roof space is needed to avoid or control condensation.

Roof design and rain protection

The design of the roof will need to take rain protection as well as the collection and disposal of rainwater into consideration.

In areas where rainfall is heavy, pitched roofs with large overhanging eaves will allow water to be shed directly to the ground and prevent back-splashing onto walls. In areas where the rainfall is less than 200 mm per annum, simple rainwater chutes are the easiest way of shedding rain from roofs.

In dry, sandy areas more sand will be deposited on the roof than rain. A rainwater disposal system that does not become blocked by sand is essential.

Design principles: roofs

Hot dry zones

- A variety of roof forms and construction are appropriate for hot dry zones. The selection may be related to functional requirements, for example a composite roof for construction of ventilation requirements and rounded and pitched roofs for thermal requirements, i.e. to reduce heat gains during the day and increase the heat release at night.

- When composite roofs are used, the roof should be of lightweight material and the ceiling of dense material. These can even be separated into two independent roof structures.
- Ventilation openings in roof spaces should be designed to remove hot air that would otherwise be trapped and transferred internally.
- To cool the external surface of the roof, its slope should be orientated towards prevailing breezes.
- Roofs sloping towards a courtyard space will allow cool air at night to flow downwards and be used to cool internal spaces.

Composite roofs can be separated into two independent roof structures. In these examples of buildings from Ouagadougou, Burkina Faso (left) architect: Hansgeorg Berger and (right) from Conakry, Guinea, the upper roofs are lightweight and the lower roofs are massive. The separation ensures that heat can be stored in the massive part of structure, but at the same time that the cooling air movement is maximised and that heat radiation impact is eliminated or reduced by shading.

(bottom left) An example from Iran of vaulted and domed mud brick roof structures that provide a thermally efficient roof. The roofs can also be used for pedestrians.

In warm humid zones the roof functions as an umbrella. It should be lightweight with large overhanging eaves to ensure that heat gains are minimised, and that rainwater is readily shed. Two separated roof structures can be used; both should be lightweight. Photos from (left) a private residence in Vaise, France, architects: Jourda & Perraudin, Lyon, and (right) a teahouse in Shugaquin Park in Tokyo, Japan.

Warm humid zones

- The roof's main function is to act as an umbrella, providing shade and rain protection.
- Large roof overhangs will protect walls and openings from heat gains and avoid rainwater splashing back onto walls.
- Pitched roofs provide rain protection and rainwater removal.
- The roof should be designed to avoid being an additional source of heat gain; this can be achieved by using composite roofs. In this case the roof and ceiling should be made of lightweight materials. Air that passes through the roof space should not be allowed to enter occupied spaces.

Wall design

Walls have several functions. Besides being structural elements, they provide protection from heat, rain, wind, dust and light, and serve as a means of space definition and partition.

Generally, external walls are exposed to considerably less solar radiation than roofs, and the period of direct solar radiation gains is shorter. The degree to which a wall is exposed to solar radiation will depend on its orientation.

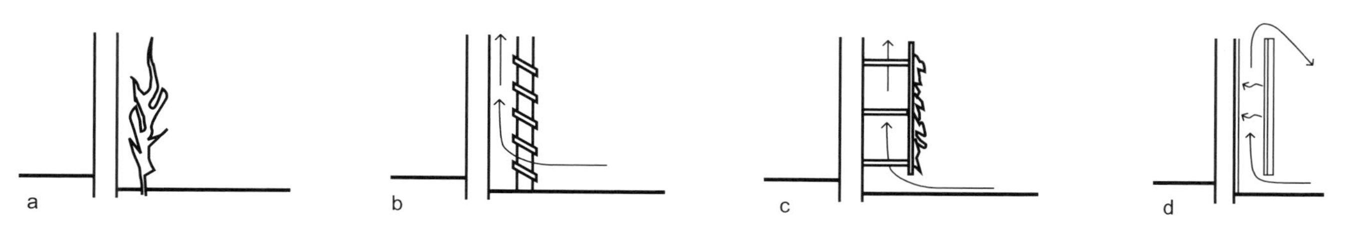

(top) Examples of wall shading devices. a. Vegetation only. b. Louvres of wood or metal. c. Vegetation on a mesh of steel or hardwood. d. An outer skin with low-emission layer toward the gap and a heat-reflective layer on the wall. The examples can be used in both hot dry and warm humid climates with the exception of a. Vegetation in hot humid regions will have a tendency to maintain humidity and cause the wall surface to deteriorate.

The entire wall can have shading devices to reduce direct and indirect solar radiation. (left) Shows an example from a church building in Berlin, Germany, with fine metal louvres. Architect: Axel Schultes. (bottom left) Shows a bank building covered with an open brick wall in Maputo, Mozambique. The wall was constructed during a rehabilitation of the building. The architect, José Forjaz, claims that the wall was not only constructed for thermal reasons, it was also a way to hide an ugly building.

In hot dry zones, walls should be massive and surfaces designed to reduce heat gains. (top) Building in Morocco. (above) Building in Bobo-Dioullasso, Burkina Faso. Architects: GET-BF assisted by the author. Engineers: BETICO.

In the northern hemisphere, north-facing walls receive little direct solar radiation and can be completely protected with shading devices for walls and openings. South-facing walls will receive a greater intensity of solar radiation, but it is also possible to shade them. The reverse will occur in the southern hemisphere. East- and west-facing walls will receive the greatest intensity of direct solar radiation, but only for part of the day: eastern walls in the morning and western walls in the afternoon.

Design principles: walls

Hot dry zones

- Generally, east- and west-facing walls should be minimised, as they will experience the greatest exposure to solar radiation.
- East-facing walls cause the most problems, as they will experience heat gains in the morning. Measures have to be taken to avoid the premature release of this heat internally, i.e. in the early afternoon, when external air temperatures are at their highest.
- Heat gains made by east-facing walls can be reduced by increasing their insulative properties, thermal mass and shading provisions, and through planning decisions such as functional zoning so that infrequently occupied spaces create thermal barriers.
- In principle, walls surrounding rooms used only during the day should be massive, whereas walls of rooms that are used only during the night should be of lightweight construction.
- In regions with large diurnal temperatures and night temperatures below the comfort level, inner and outer walls need thermal mass to balance temperature variations.
- In regions with rather low diurnal temperature ranges, thermal mass should be restricted to internal walls, partitions and intermediate floors.
- Operable ventilation openings should be included in walls for night-time ventilation requirements.

Warm humid zones

- In warm humid regions, walls should in general be of lightweight construction. This is critical to reduce the amount of heat stored by the walls.
- Walls should be designed to maximise internal air movements while protecting against internal heat gains and external radiation.
- Solid walls should be minimised. Openness should be the dominant structural feature. Walls can take the form of adjustable shading devices.
- Maximum solar radiation will fall on east- and west-facing walls, and these should be kept as short as possible.
- Since humidity levels are higher in the mornings than in the afternoons, and since the sky tends to remain overcast until midday, west-facing walls will receive more solar radiation than east-facing walls. West-facing walls are the most critical.
- If walls are exposed to direct solar radiation, such as gables, parapet walls, east and west walls, the addition of mass or insulation can prevent the increase of internal temperatures. The release of stored heat must be managed so as not to increase internal temperature. In general, thermal mass should

In warm humid zones walls should be light and transparent in order to continuously ventilate internal spaces and to reduce heat storage. These demands are catered for in traditional houses in Japan, shown in the photo (above). The double wall with a corridor between assisted by different sliding wall elements will be flexible enough to control ventilation and light as well as heat and cold. (left) A flexible open wall structure with blinds and shutters that can cater for controlled cross-ventilation for different wind direction. The building also has a large roof overhang. Alderton house, Education Centre in N.T., Australia. Architect: Glenn Murcutt.

These houses from a compound in Darwin, Australia, have walls and openings well protected by means of balconies, projected roofs, sliding shading devices and vegetation. Even though it is a rather high-density compound, natural ventilation still seems possible owing to a variation in forms of structures and to the various projections, which will catch the wind. Troppo Architects, Drawin.

be restricted to construction elements that are not exposed to solar radiation, such as internal walls and partitions.

- The outer surface of walls should be light in colour to reduce the transfer of heat.

Notes: Special design

A building like the one in these photos may be found comfortable in hot humid climates, contrary to what has been said in the text. It is a private residence in Havana, Cuba, which has a ground floor with solid walls. It has a first floor made of wood, i.e. non-heat-absorbing material, and a courtyard with dimensions to ensure air movements, but small enough not to expose the heavy walls to much direct sun during the hot season. Most rooms can be cross-ventilated. The extreme thickness/density of the wall on the ground floor may absorb the heat encountered during the day. Studies made by B. Givoni, mentioned in his book *Climatic Considerations in Buildings and Urban Design*, reveal that such a building may, owing to its capacity to store heat, be able to store the 'heat of day', but only very effective night cooling by mechanical means can ensure continuous comfort during the night.

FLOOR DESIGN

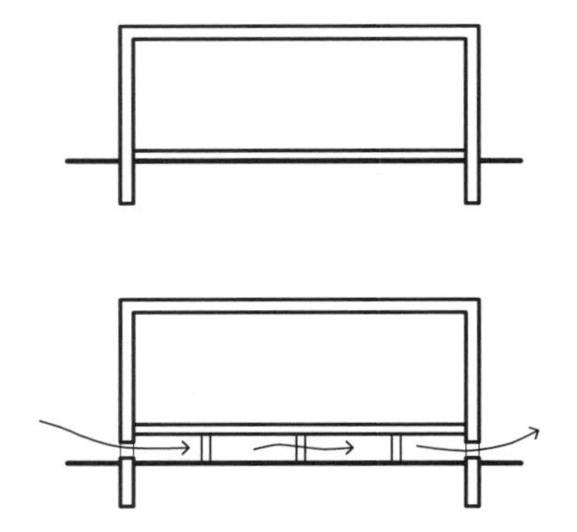

When heat storage is required the ground floor should have maximum contact with the ground. When maximum air movement is required to reduce heat gains, the ground floor should be elevated.

(right) The cooling potential offered by the floor is well understood in hot dry zones, where the floor is often used for sitting. From Mit Rehan, private residence in Sakkara, Egypt. Architect: Hassan Fathy.

The ground floor of buildings can reduce heat gains by being designed to:

- increase its overall thermal storage capacity;
- increase its overall ventilation potential.

The first point involves a building maintaining maximum contact with the ground and the second involves elevating a building off the ground.

The ground has valuable heat storage capabilities. Heat will be conducted from a building's envelope to the ground. In areas where thermal storage is required, i.e. where daytime temperatures are considerably higher than night-time temperatures, the floor can be used if maximum contact with the ground is maintained. The floor is normally constructed as a slab on the ground.

In areas where maximum ventilation is required to reduce heat gains (where a building cannot cool down sufficiently during the night to allow heat storage during the day), the floor can be elevated off the ground to allow the cooling effects of air movement to be utilised. The floor in this case should be of lightweight construction to reduce heat storage possibilities.

Design principles: floors

Hot dry zones

- In hot dry zones a floor that stores heat is required.
- Ground floors should be solid and laid in direct contact with the ground, to utilise its thermal properties.
- The floor should be of heavyweight construction, for example stone, clay or a concrete slab on the ground.
- In hot dry areas, it is common to sit on cushions and carpets on the floor. This shows a cultural (and thermal) awareness of the cooling potential of floors.
- To take further advantage of the ground's thermal properties, buildings can include cellars or be built completely underground (see the section in Chapter 7, Earth cooling).

Warm humid zones

- Ground floors should have no contact with the ground. This is to ensure that the floor does not store heat.
- Raising the floor off the ground will provide better exposure to prevailing breezes and allow air to be cooled as it passes over vegetation below the building and thereby cool the floor. It should be raised only about 50–60 cm to profit from the cooling from the vegetation. If, however, the wind exposure is found more crucial, or maybe for functional reasons, the floor could be increased to a full storey height.
- An elevated floor is advantageous in reducing impact due to heavy rains. It offers protection from rotting and termites.
- The floor should be made of lightweight construction (e.g. a single layer of timber boarding on floor joists).
- The provision of gaps between floorboards will improve ventilation further. Careful detailing is required to provide protection from the entry of insects and to reduce noise transmissions for privacy reasons.
- If a solid floor is needed it could have built-in ducts such as a hollow concrete slab with the double purpose of reduction of thermal storage capacity and providing ventilation with natural cooling.

The photo is an illustration of the separation of daylighting and ventilation. The top windows under the ceiling create a well-distributed indirect daylight whereas the shutter below is used during hot summers to ventilate and cool the body and give a pleasant view. From architect Alvar Aalto's drawing office, Helsinki, Finland.

Opening design

The appropriate design of openings is critical as it is at these points that the building envelope links the internal environment directly to external conditions.

During the day openings will admit solar radiation, which consists of heat and direct or indirect sunlight required for lighting. They will also admit external air required for ventilation purposes, which will facilitate the removal of internal heat gains from a building.

Design choices as regards the size, location and treatment of openings will affect heat gains, air movements and the quality of light and air.

Distinctions can be made regarding an opening's functions. They can be designed to function only as air inlets or outlets for ventilation purposes, or to permit only sunlight for daylighting purposes. The needs for positioning and size of openings for daylight as opposed to ventilation openings are generally conflicting, and it is advisable to separate these functions in different openings. However, sophisticated solutions can combine several functions in a single opening.

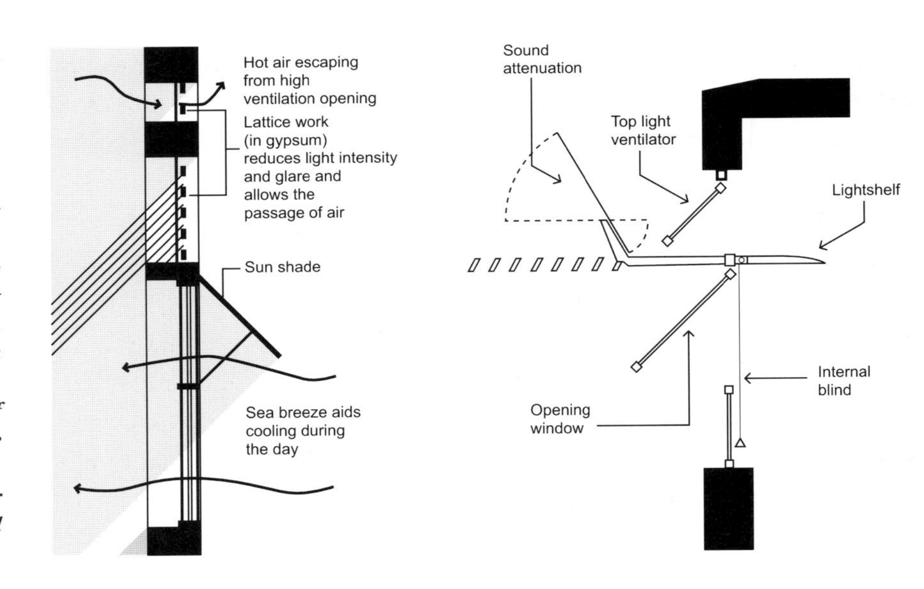

Examples of composite solutions for openings:
(left) A traditional example from the Middle East, showing the shading functions and the ventilation system in a combined design. Based on a drawing by Development Workshop.
(right) A contemporary example of a single opening designed to serve several functions: sun protection, ventilation, reflection of light for better distribution of daylight, and the reduction of noise transmission.

Opening design for daylighting

External conditions and internal demands for specific quantity and quality levels will govern the design of daylight openings, unlike artificial lighting, which can be designed independently of location, climate or even the building envelope.

The location, size and treatment of openings will have an effect on the quantity and quality of light entering a building and the amount of heat gain and heat excluded from a building.

An answer to increased daylight through an opening by a light-reflecting wing wall. From a church building in Klippan, Sweden.

Sources of light

Sources of daylight into a building are as follows:

- direct light along a straight path from the sun;
- diffused light from the sky;
- externally reflected light from the ground or other buildings; and
- internally reflected light from internal walls, ceilings or other surfaces.

Sources of glare

Glare results from strong contrasts in light levels or high levels of light. Bright light that is reflected off light-coloured surrounding surfaces will be a source of glare. It is important that openings are protected from external glare sources as glare can create a very uncomfortable visual environment.

Daylighting and climate characteristics

Hot dry regions are characterised by sunlight from cloudless skies. However, the sky's luminance may be as low as 1700 cd/m², which may not be enough to ensure adequate daylighting. Often light dust is suspended in the air, which creates a haze and increases the apparent sky brightness up to 10 000 cd/m². Heavy, dusty sand storms can reduce it to below 850 cd/m². A clear sky usually has the highest luminance near the horizon.

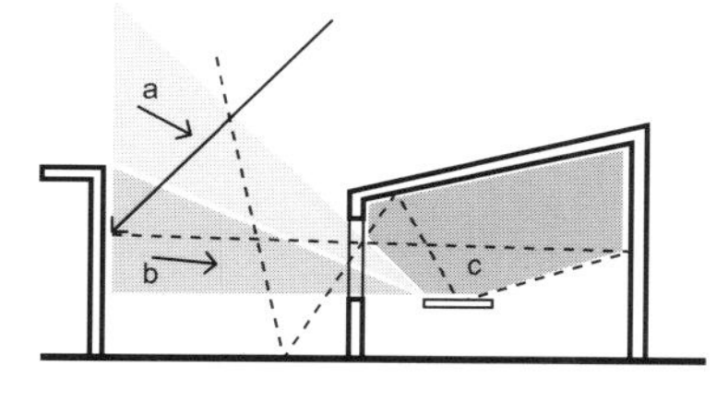

Sources of light entering a building.

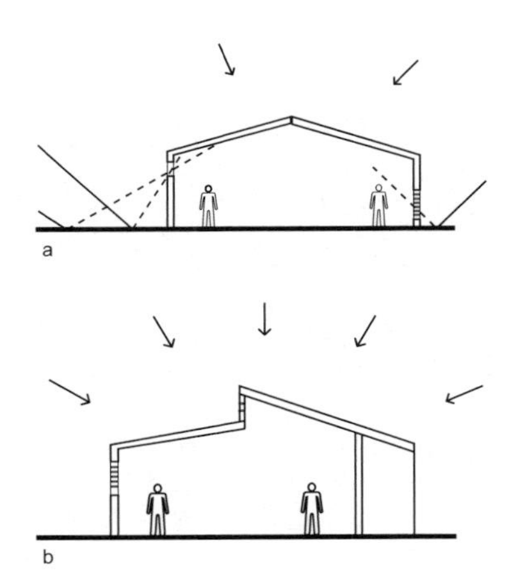

a. In hot dry zones glare from reflecting surfaces and the horizon can be controlled by indirect light and/or carefully positioned and protected openings.
b. Glare from the sky in warm humid zones can be controlled using overhanging roofs, louvres and screens.

Openings positioned high in walls will improve internal light distribution and reduce direct light/ heat. From a building in Yazd, Iran. An appropriate sollution for hot dry zones.

In warm humid regions, the sky is typically overcast, with a luminance often exceeding 7000 cd/m^2. The proportion of diffused light or skylight is predominant, and a very bright sky, viewed from a moderately lit room, can be a source of discomfort due to glare.

Note: 1 candela (1 cd) is approximately the intensity of light on a horizontal plane from one ordinary candle.

DESIGN PRINCIPLES: OPENINGS FOR DAYLIGHTING

Hot dry zones

- Direct sunlight must be avoided for thermal reasons, and openings should be as small as possible.
- Openings should be positioned so that the main view is directed towards the sky, rather than the horizon or ground. Reflected light from the ground and other light-coloured surfaces is often a source of glare.
- Diffuse internal light can be achieved by screening, e.g. Mashrabia, or by openings towards a shady and green courtyard to avoid glare and gain privacy.
- Maximum use of indirect and internally reflected light is the most appropriate form of daylighting. For instance, a high-level opening reflects light towards the ceiling. If the ceiling is painted white, well-diffused interior lighting can be achieved even through relatively small openings.
- Low-level openings are acceptable if they open onto shaded, green areas or non-glare surfaces.
- Care must be taken that shading devices do not create glare sources.

Warm humid zones

- Because the sky is the main source of glare, openings should not be positioned to face the sky.
- As large openings are needed for cross-ventilation, low overhanging roofs or wide verandas can be used to obstruct the view of the sky.
- Shading devices can be used to exclude unwanted light. Louvred or adjustable shading devices are effective if designed to reflect light from the ground up towards the ceiling.

Adjustable louvres can be positioned to reflect light from the ground and exclude unwanted sky light. A building from Kemsey, Australia. Architect: Glenn Murcutt. An appropriate sollution for warm humid zones.

Notes: Supplementary artificial lighting

It will not always be possible for daylighting alone to satisfy all of a building's internal lighting requirements. In some cases the function of the building or part of a building will require controlled lighting levels. Alternatively, a building that is occupied during the night will have no alternative but to rely on artificial light. However, choices can be made to use energy-efficient light fixtures or flexible modes of lighting, depending on the quantity and quality of light required. For instance, task-related lighting could allow occupants to control the use of their specific light source rather than have a blanket of lights switched on. Artificial lighting consumes energy, and creates a heat output that needs to be considered.

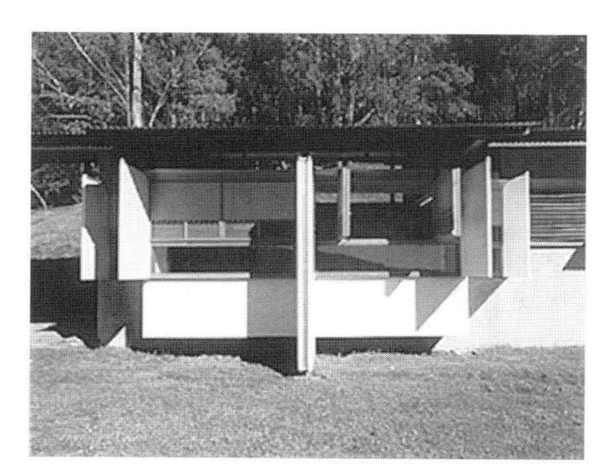

A multifunctional opening design for light and wind control, i.e. blinds, shutters, light and wind-reflecting or protecting sidewall. An educational centre in Riversdale, NT, Australia. Architect: Glenn Murcutt.

Opening design for ventilation

Ventilation is supposed to provide fresh air in appropriate volumes and speed of air flow to facilitate, for example, evaporative cooling of the air and the occupants as well as cooling of the structures. The appropriate location and size of openings for ventilation seldom correspond to the appropriate location and size of openings required for lighting, and the two functions should be treated separately even though it might be found feasible to combine them in some situations, e.g. in warm humid regions where there is a need for a great deal of air movement through the walls which at the same time could produce sufficient daylight.

Natural ventilation can occur in two ways:

- air movement due to wind-generated pressure differences; and
- air movement due to temperature-generated pressure differences.

Below is a brief summary of the basic way in which openings influence air movement patterns and speeds through buildings. (A more detailed explanation of ventilation is provided in Chapter 7.)

Wind-generated pressure differences

Air movement will generally occur if pressure differences exist across a building. Openings can be designed to use these pressure differences.

The location, size and treatment of openings will influence the quantity and quality of external air drawn into cool internal spaces.

The size of openings on the windward and leeward sides of a building will affect air flow internally. A relatively large opening on the windward side causes lower air velocities compared with a smaller opening facing the wind. Small openings on the windward side and large openings on the leeward side will create higher air velocities and more changes of air, but also a poorer distribution of air throughout the room.

Differences in height between leeward and windward openings will also create variations in internal air distribution.

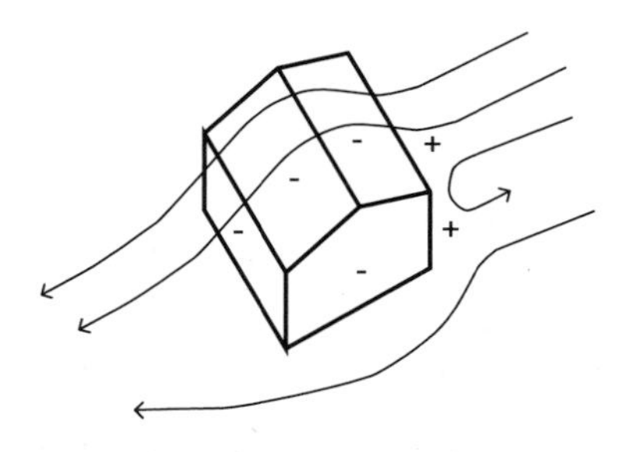

Wind-generated pressure difference across a building.

Openings in opposite walls will produce different internal air distribution patterns from those occurring as a result of openings located in adjacent walls.

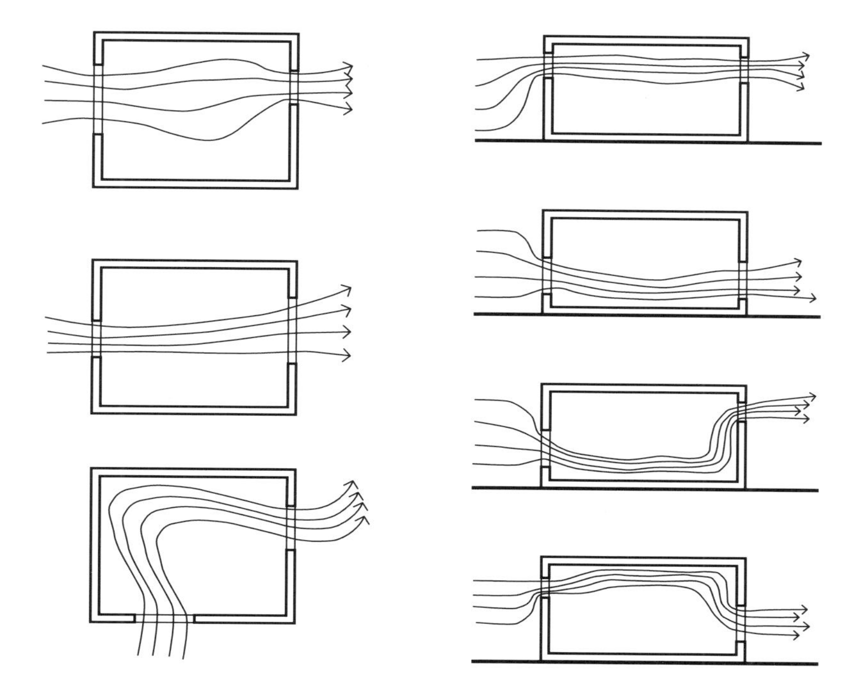

(far left) Variations in air distribution and speed due to differences in the size and positioning of air inlet and outlet in walls, shown as horizontal sections.

(left) Variations in air distribution and speed due to differences in the size and positioning of air inlet and outlet in walls, shown as vertical sections.

Temperature-generated pressure differences

Natural ventilation is influenced mainly by wind-generated pressure differences across the outside of the building. However, without the influence of wind pressure differences, if openings are located at different levels in two zones, e.g. two openings inside or one outside and one inside, with a temperature difference between the two zones, there will be an air movement. It is the natural tendency of warm air to rise and try to escape through the top opening. This phenomenon is referred to as the stack effect. Another natural phenomenon, called buoyancy, assists the stack effect as a ventilation agent. Cool, dense air settles at the lower part of a space, and warm, less dense air rises to the upper part of the same space.

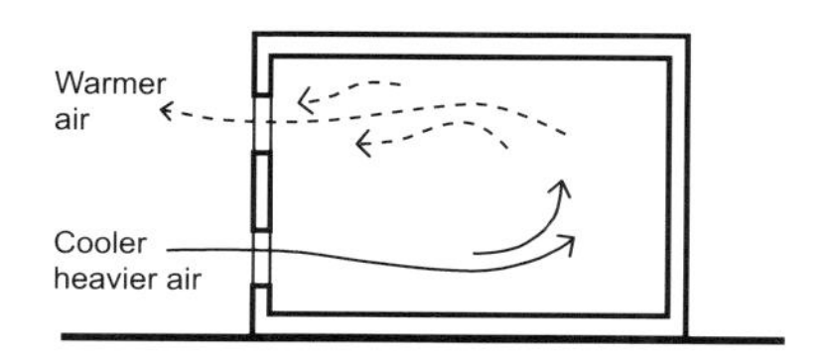

The movement of air due to temperature-generated pressure differences across openings.

When wind speeds are low or external air temperatures are lower than internal air temperatures, i.e. at night, the stack effect

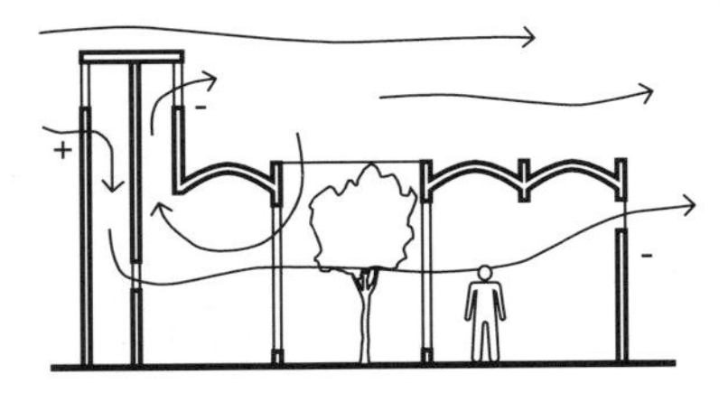

Ventilation must be controlled due to undesirable ambient air temperature and dust. The stack and suction effect can be utilised to generate ventilation, if openings are located at differing heights.

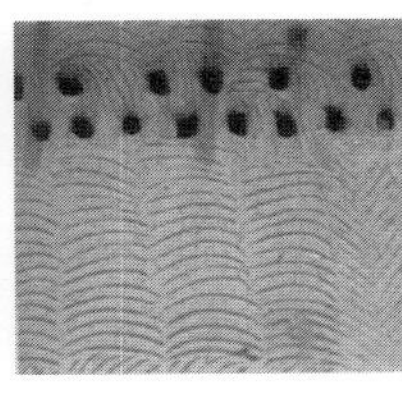

Small simple openings located high in walls can ensure minimum air exchange without increasing heat gains considerably. They can be controlled by internal shutters. During the night they will allow cool air to enter and pass over the warm underside of the roof.

can be used to generate air movement that can create ventilation and cool down the structure at night.

Design principles: openings for ventilation

Hot dry zones

- Openings for ventilation and daylighting purposes should preferably be separated and function independently.
- Ventilation openings need to be operable to address diurnal temperature variations. High daytime temperatures will require that ventilation openings be closed during most of the day to reduce heat gains.
- To profit from cooler night-time temperatures, the ventilation openings should be opened to aid the removal of stored heat in the structure.
- In order to use the stack effect to cool down a structure, openings for ventilation should be located at different heights (vertical ventilation).
- Ventilation openings in walls and roofs should be located to allow air to move over the warmest internal surfaces, e.g. the ceiling.

Ventilation openings just below the roof level as part of the ventilation system in an internal street. They also serve as a light intake/reflection device. The Botswana Technology Centre, Gaborone. For reference see page 67.

Warm humid zones

- Because of relatively small differences in temperature during the day and night, continuous wind-generated ventilation of internal spaces is required (horizontal ventilation), except during the midday hours in the hot season.
- All openings will need to be as large as possible to maximise any external air movement.
- Openings should be placed on opposing facades to facilitate cross-ventilation of rooms. Large openings should be placed not only in external walls but also in internal walls.
- Openings should be located to allow air movement to occur at body level. This will aid in evaporative cooling of the body.
- Openings can be designed to utilise winds and/or the stack effect.

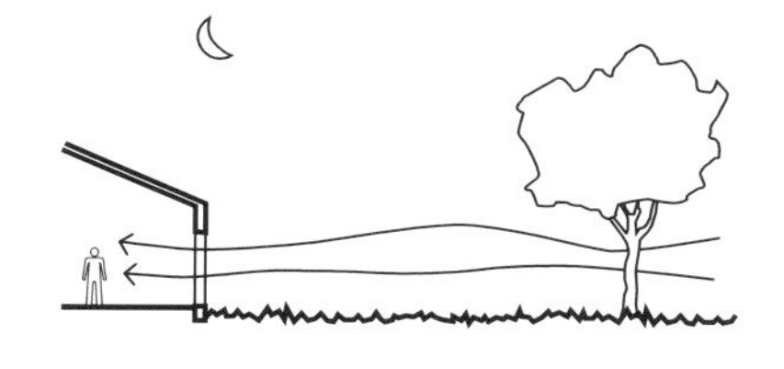

Continuous ventilation of internal spaces is required most of the year and the day. Openings should be located to direct air movement across occupancy level and to cool the structure.

Openings should be as large as possible and located to maximise cross-ventilation of internal spaces. (left) A rainforest research and education centre in Endaun Rompin, Malaysia.

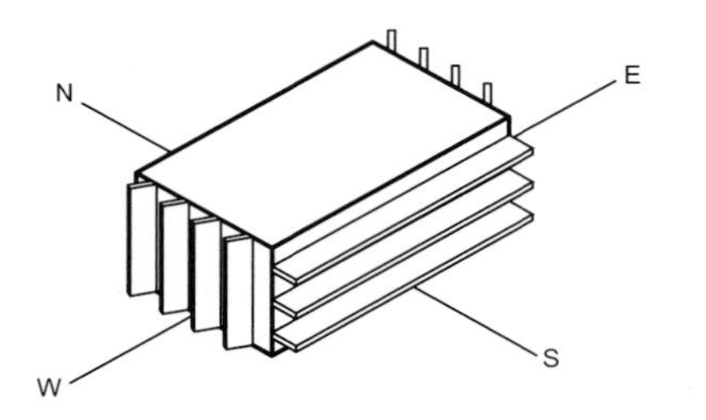

The two basic forms of shading device (horizontal and vertical) that should be adopted for different orientations. Shown for the northern hemisphere.

DESIGN OF SHADING DEVICES

Various options can be used in the treatment of openings to exclude or reduce solar radiation and heat gains. These can include proper orientation, sizing and location as well as building envelope related measures such as wall protection, colonades, balconies and roof overhang. However, the most common and effective form of protection is the use of shading devices.

Shading devices will not only exclude or reduce solar radiation, they can also be used to reduce glare and provide protection from the rain. They can influence internal air movement patterns (as discussed in Chapter 7). The material used to construct a shading device will have an effect on what is normally called the solar protection factor. An opaque material will have a solar protection factor of 100%, whereas all translucent or transparent materials such as fabric, screens and vegetation will have less. Solar protection devices should not only cater for direct solar radiation, but also diffuse or reflected solar radiation must be considered. This is particularly important for buildings in warm humid climates, where a great deal of the solar radiation is diffuse.

It is vital not to adopt a universal shading system for all the openings of a building. The choice of shading devices should reflect both orientation and prevailing climatic conditions.

There are two basic forms of shading devices, as follows.

- If the sun passes high in the sky across an opening, a horizontal shading device can be used to exclude solar radiation. This is effective for north- and south-facing openings.
- If the sun passes low in the sky to shine into an opening, a vertical shading device can be used to exclude solar radiation. This is effective for east- and west-facing openings.

DESIGN PRINCIPLES: SHADING DEVICES

Regardless of the type of shading device used, it should:

- be placed on the outside of an opening;
- be made of light and reflective materials, to avoid absorption and re-radiation of the heat through an opening;

- be made of materials with low heat storage capacity (i.e. to ensure rapid cooling);
- be designed to prevent reflection onto any part of the building or openings;
- be designed so that hot air is not trapped.

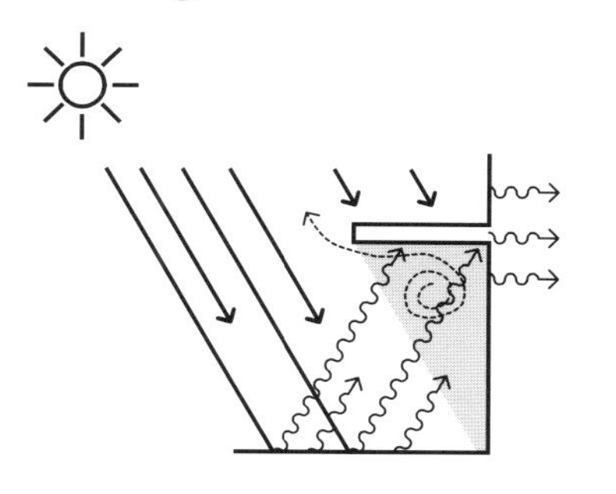

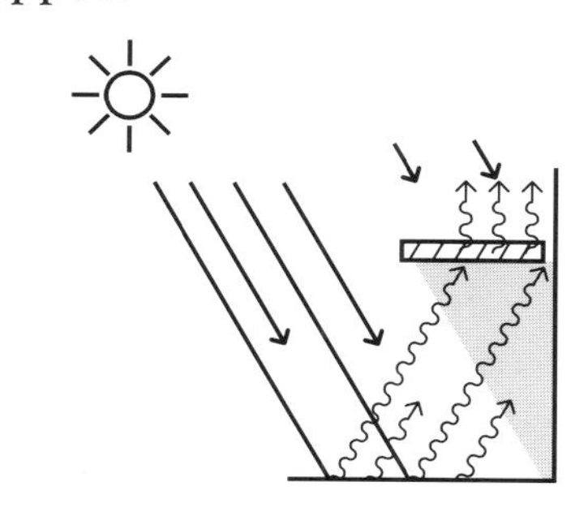

(left) A shading device can increase the impact of heat on a building. It can reflect heat onto the building and it can also trap hot air and thereby cause heat to be transferred inwards through the structure. (right) This can be avoided, for example, by separating the shading device from the building, by using light, reflective materials, and by using open e.g. louvred shading devices.

Examples of shading devices

(left) Horizontal and vertical shading device on a building in Sydney, Australia.

(right) Combined vertical and horizontal shading device (of the wall and the openings) on a building in Ouagadougou, Burkina Faso.

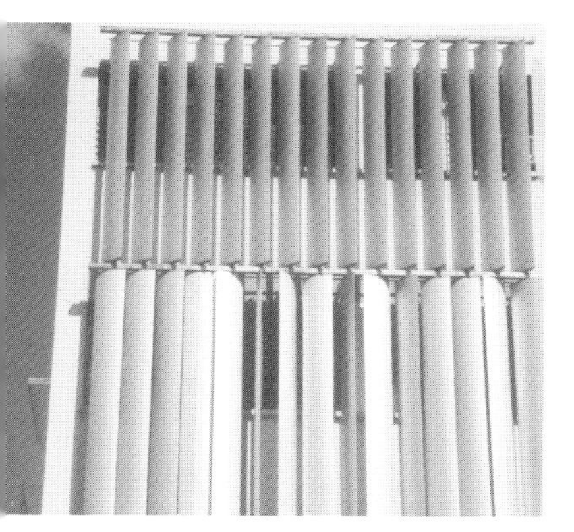

(left) Vertical shading device on a building in Basse-Terre, Guadeloupe.

(right) Horizontal shading device on a building in Maputo, Mozambique. Architect: Jose Forjaz.

(left) Transparant horizontal shading device on a building in Lyon, France. Architects: Jourda & Perraudin.

(right) Transparant vertical shading device on a building in Frankfurt, Germany.

(left and right) Horizontal and vertical shading devices as aluminium grating with tilted internal wall to compensate for the position of the sun. The Botswana Technology Centre, Gaborone. For reference see page 67.

Shading devices on a building in Herfordshire, UK. Architect: Nicholas Crimshaw & Partners Ltd. Engineers: Ove Arup & Partners.

6. Thermal properties of building materials and elements

The effectiveness of a building envelope is directly related to the choices made regarding materials and construction, that is, the thermal properties that each material has, and the properties they have when combined together to form building elements.

One may make thermal decisions to achieve the greatest thermal potential from materials or elements only to find that these may conflict with earlier design decisions. This is particularly critical in an active design process as every choice made will have its own set of causes and effects.

The manner in which a material will react to thermal forces will be determined by its thermal properties. This term refers to a material's composition and characteristics, and the way these qualities react under different thermal conditions. This will also govern the manner in which a material will respond, that is, how it:

- *reflects heat*
- *absorbs heat*
- *emits heat*
- *reduces the flow of heat*
- *stores heat and releases it.*

The first three processes will occur at the surface of a material, involving heat gains or losses due to radiation. These can be manipulated to control heat entry at the surface of the material. (Translucent or transparent materials not only reflect, absorb and emit but also transmit solar radiation. See the section on Design principles: thermal properties for openings at the end of this chapter.)

The last two processes involve a heat exchange process, whereby heat gained on the exposed surface of a material is transferred to the inside surface of a material. These can be manipulated to control heat transferred through a material.

Materials and the manner in which they are combined to form a building element will always have a combination of these properties. The choice of the materials for the building envelope in the climatic zones in this Guide is normally made to avoid unwanted heat radiation towards the interior.

Studies of thermal properties of materials and building elements from the technical literature indicate quite a lot of variations in these properties. This could be due to differences in or lack of a precise definition of materials and building elements. The values in the tables below are compiled from various sources and should be considered as indicative only. They should serve only as a basis for comparison of different materials and building elements.

The quality of the execution of construction has an important effect on the thermal values of building elements and makes any thermal simulation 'a good guess', the quality of which depends entirely on the quality of information available and the experience and expertise of the person doing it. This is particularly true for hot dry and warm humid climates owing to the variations of temperatures and humidity throughout the day and over periods of the year. The application of passive measures for ventilation and cooling only adds to the complexity of the issue. This could be one of the reasons for the use and popularity of active measures, which will ensure at least one stable or controllable element in the analysis and in the everyday use of the building.

Thermal effects on materials

A material will experience the effects of thermal forces through radiation, convection and conduction.

- Radiation will impact on the exposed surfaces of materials either as direct solar radiation from the sun or as a radiant heat exchange with its surrounding environment (i.e. the ground, other materials, buildings, etc.) and the sky in the form of long-wave radiation. The Earth's atmosphere is also responsible for diffuse solar radiation, which will also impact on materials.
- Convection is the heat transfer by air movement to a material. It will occur when there are temperature differences between the surrounding air and the temperature at the surface of a material.
- Conduction is the process of heat transfer through a solid material from the hot side to the cool side. The rate of this transfer will depend on the conductive properties of the material and the temperature difference at which this occurs.

Materials will experience a heat input (or gain) when surrounding objects are of a higher temperature, which occurs generally during the day. Materials will experience a heat output (or loss) when surrounding objects are of a lower temperature, which more often occurs at night.

Controlling heat entry

Reflectivity properties

The first line of heat control or heat entrance for an opaque material or building element will occur at its surface. Generally, the surface temperature of a material or element will be higher than that of the air temperature when it is exposed to solar radiation. This will result in heat being absorbed (i.e. gained) by the surface and being transferred through the material or element.

However, some materials have properties that will reflect heat rather than absorb it. This thermal property of a material or element is referred to as its reflectivity.

Reflectivity will determine the amount of radiation that is not gained by the surface. The reflectivity properties of a surface are generally associated with colour. Lighter-coloured surfaces reflect radiation and heat gains, whereas darker-coloured surfaces reflect poorly and allow more heat gains.

The importance of reflectivity is that it can be used as the first line of defence against heat entry as it directly stops the heat from being absorbed on a surface. Below is a list of some common materials and surfaces and their reflectivity properties. It shows the amount of radiation that will be given back by a surface to its external surrounding owing to reflectivity. Reflectivity is expressed as the percentage of the incident radiation that will be reflected from a surface.

Examples of average reflectivity

Materials			Reflectivity (%)
Brick and concrete, light			40
Brick and concrete, dark			20
Brick, white glazed			75
Concrete, smooth			35
Concrete, rough			20
Tiles, white glazed			80
Aluminium, shiny			90
Asphalt			10
Painted surfaces			
Glossy white or whitewash			70–90
Off-white			65
Black			5
Other colours:	Light	Medium	Dark
Yellow	70	50	30
Brown	55	25	10
Blue	60	20	10

(cont'd...)

(...cont'd)	
Soil and vegetation	
Sandy soil	25
Light, dry sand	35
Dark, cultivated moist soil	10
Grass	25
Dry grass, straw	40
Vegetation, dark	15
Vegetation, light	25

(From various sources.)

Absorptivity and emissivity properties

If radiation on an opaque surface is not reflected, it is transferred to the surface, i.e. the heat impacting on a surface is gained by the surface. This property is referred to as a material's or element's absorptivity.

The surface of a material will absorb as well as emit radiation. Absorption involves heat gains whereas emittance involves heat losses or releases occurring from a surface. This is referred to as the emissivity of a material. For a given wavelength of light the absorptivity and emissivity are the same. For a building surface the absorptivity from solar radiation (short-wave radiation) has a certain value. This value is different from the emissivity value, which is the heat radiation (long-wave) from the surface to cooler environments – which takes place at night in hot climates.

Reflectivity as well as absorptivity and emissivity properties of materials are important means of controlling the impact that radiation will have on an exposed surface. Surfaces and/or materials that reflect rather than absorb radiation, and which can emit radiation, will minimise heat impact.

A heat bridge may occur for shading devices that are supposed to receive radiation, and to protect the building from direct radiation. When the shading device is made of heat-absorbing and not reflective material, it will act as an added heat load on the adjacent building element.

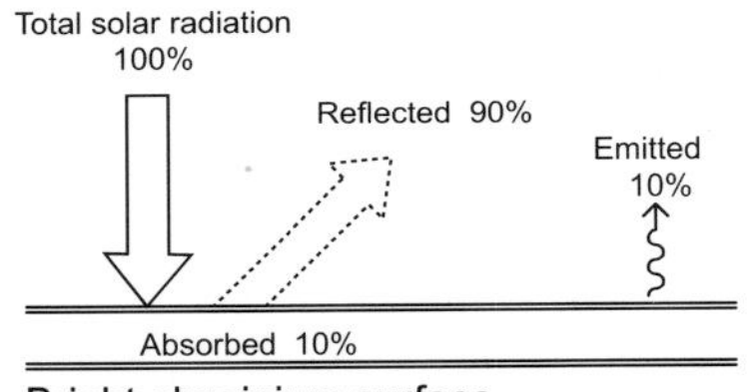

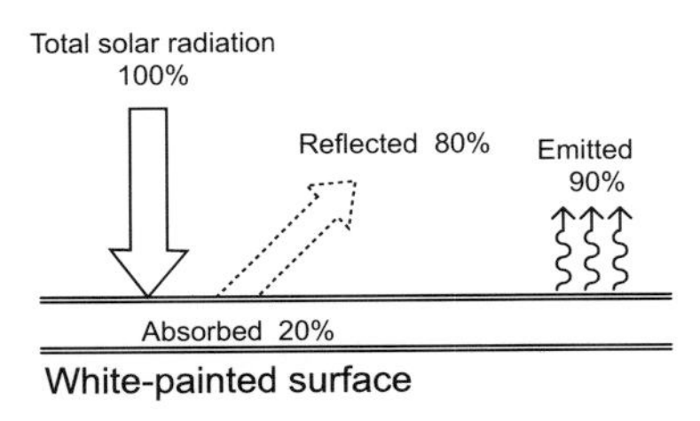

An example of the surface behaviour of two different materials.

Below is a list of materials and building elements and their absorptivity and emissivity properties. The absorptivity is an expression of the short-wave radiation that is gained as a percentage of the total amount of radiation impacting on a surface, e.g. between 100 and 400 W/m² depending on location, the time of day, cloud cover, etc. for a whitewashed horizontal surface. Emissivity is an expression of the long-wave/heat radiation that is lost as a percentage of the maximum radiation possible from a non-metallic and mat surface. For example, the net radiant heat loss at night for a whitewashed surface will be approximately 160 W/m² with a clear sky and a roof temperature of 27°C.

Surface conditions will also influence the amount of heat gained or lost.

Examples of surface absorptivity and emissivity properties in %

Surface	Condition	Absorptivity (short-wave)	Emissivity (long-wave)
Aluminium roofing sheet	New	10–25	10–25
	Oxidised	30–50	20–50
Aluminium paint		40–50	40–55
Galvanised roofing sheet	New	20–30	20–30
	Rusty	60–85	70–90
Fibro cement sheets	New	35–50	85–95
	Blackened by algae	60–85	85–95
Aluminium foil	New and bright	5–10	5–10
	With dust film	10–20	10–20
Whitewash		10–15	85–90
Light-coloured brick or plaster	New	20–30	85–90
	Weathered	30–40	85–95
Red brick, stone or tile		65–80	85–95
Concrete tile or slab		45–65	85–95
Asphalt		75–90	85–95
Vegetation		5–10	85–95
Soil		15–25	85–95
Sand		30–40	85–95

(from various sources)

Air gaps or infiltration

Poor design detailing and construction will result in the creation of air gaps or infiltration through which heat can enter – that is, the entry of uncontrollable or unpredicted heat. Here are just a few examples of where air gaps may occur:

- around doors and windows
- windows and glass fittings
- plumbing pipes and electrical conduits passing through walls
- the joints between the ceiling and the walls
- the roof and wall
- fixtures placed in walls (e.g. exhaust fans).

Controlling heat transfer

When radiation penetrates the surface of a material, a heat exchange process takes place owing to temperature differences between the outside air and the surface. The way a material or building element will react when heat is transferred occurs in two ways:

- The material may interfere with the transfer of the heat from its outside surface to its inside surface and thereby reduce the amount of heat that is gained internally. This is referred to as a material's insulative properties.
- The material may store the heat gained from the outside and thereby delay the transfer of this heat internally. This is referred to as a material's thermal storage and time lag properties.

These properties will have considerable consequences for the internal thermal environment, as the amount of heat that is transferred will affect internal temperatures.

Insulative properties

The insulative property of a material or a building element is based on the thermal conductivity of the material, the λ(lambda)-value, which is relatively easy to examine. It is the rate of heat flow through a specific volume of a material with a unit temperature difference between the two sides. The λ-value is measured in W/m °C and is an expression of a material's capacity to conduct or transfer heat as opposed to its capacity to resist heat flow, which is the insulative property. There is a reciprocal relationship between the capacity to conduct heat and to insulate. In general, high-density materials can transfer more heat than lightweight materials and therefore have a higher λ-value. Lightweight materials insulate well owing to their high percentage of air, which is a good insulator with a low λ-value. The table on the next page indicates the density and λ-value for different dry materials. The λ-value increases with humidity.

Air cavities

Air is one of the best insulators. Materials and building elements that enclose, trap or contain air have low heat transfer characteristics. The resistance to heat transfer of a wall or roof can be increased by the creation of air cavities between the layers of construction, as in double walls or roofs and in composite building elements.

The heat transfer that will occur across a cavity will take place by both convection and radiation. Radiation across a cavity can be reduced if the surfaces of the cavity have low absorptivity and emissivity properties, e.g. aluminium foil or shiny corrugated aluminium sheets. Whenever a cavity is incorporated as part of a composite building element, the element with the greatest mass (i.e. with high thermal storing properties) should be located on the inner side of the cavity, and the outer side should be of lightweight construction and preferably with high insulative properties. This will allow the amount of heat transferred through the element to be reduced before it reaches the layer that has heat storage properties. This will aid in the reduction of internal heat gains.

Approximate density and thermal conductivity properties of various materials

Material	Density (kg/m³)	λ-value (W/m °C)
Aluminium	2700	200
Steel	7850	60
Stone	2600–2800	2.3–3.5
Concrete:		
light	1000–1700	0.3–1.0
dense	2400	1.8
Timber:		
lightweight	400–500	0.15
heavyweight	600–800	0.20
Compressed earth block	1700–1800	1.0–1.2
Clay	1600–2840	0.45–1.8
River sand	1700–2000	1.3–1.5
Air (still air)	1.2	0–02
Water, 10°C	1000	0.58
Mineral wool and glass wool	20–120	0.03–0.05
Polystyrene	20	0.035
Natural wool fibre	200–300	0.06
Cellulose fibre	60	0.04
Peat fibre	225	0.05

(from various sources)

The insulative properties of material and building elements are expressed as the U-value (W/m² °C). U-values are calculated on the basis of material thickness and the value of 1/λ for a homogeneous material plus the surface resistance on the two surfaces. For a composite building element, the U-value is based on the sum of 1/λ for each homogeneous layer plus the surface resistance on the two outer surfaces of the building element. For composite elements, an increase in the overall U-value due to thermal bridging must be considered, i.e. concentration of flow or transfer of heat due to high density/conductivity structural elements in a composite building element, such as steel or concrete beams or columns mixed with insulation materials between an outer and inner skin.

Overleaf is a table showing examples of various materials, building elements and composite building elements and their insulative properties. Materials with low U-values have high insulative properties or resistance, and those with high U-values have low insulative or resistance properties.

U-values are not global values; they vary according to the surface properties of the material and climatic exposure, i.e. orientation, altitude, humidity, sun exposure, wind pressure, etc. This is particularly significant for complex building elements such as windows, where infiltration and thermal bridging are unavoidable. The U-value for ordinary windows exposed to sheltered and severe climatic conditions is given below. The humidity in a material itself, for example, can play an important role for the U-value. An insulation material such as rock or glass wool cannot absorb humidity, and the insulative capacity drops significantly owing to increased conductivity in the case of humidity/condensation in the material. This is contrary to, for example, natural wool and adobe structures, which can absorb and store humidity to a certain level without losing their insulative capacity.

Approximate insulative properties of various materials and construction types

Type of construction	U-value (W/m² °C)
Roofs	
Corrugated cement sheets	7.9
Corrugated iron sheets or tiles on a wooden frame construction	8.5
As above + plasterboard ceiling	3.2
Reinforced concrete slab, 100 mm, screed 63–12 mm, three layers bituminous felt	3.4
As above with insulation on the screed:	
50 mm wood wool slab	1.1
130–80 mm aerated concrete	1.4
130–80 mm foamed slag concrete	1.5
Walls	
Solid brick, unplastered 115 mm	3.6
Solid bricks, plastered both sides 230 mm	2.4
Double brick wall (2 x 115 mm), with 50 mm cavity, plastered on the inside	1.7
Double concrete block (2 x 150 mm), 50 mm cavity, outside rendered, inside plastered made of:	
Hollow concrete block	1.2
Clinker concrete blocks, massive	1.1
Unstabilised adobe block, 300 mm	0.7–1.1
Pressed stabilised soil block, 300 mm	0.4–0.8
Concrete, ordinary, dense, 200 mm	3.2
Stone, medium, porous, 300 mm	2.8
Floors	
Concrete on ground or hardcore fill	1.1
As above with wood boards finish	0.9
Timber board on joists, underfloor space ventilated, linoleum finish	1.4
Windows (ordinary wooden structure with operable window frames)	
Sheltered climatic exposure:	
single glazing	4.0
double glazing 12 mm space	2.2
Severe climatic exposure:	
single glazing	7.4
double glazing 20 mm space	3.2

(from various sources)

Insulative capacity (the U-value) of a building element is of particular importance where a difference in temperature between the two sides of the element should be maintained, e.g. an element where cooling is introduced on the inside of the element while the outside of the element is exposed to the exterior and is hot. In such a situation, and with a high-capacity layer of insulation close to the inside surface, introduced cooling will be effective by reducing air temperature and lowering the radiation from the element. If cooling cannot be introduced or can be introduced only to a limited extent, there is a need for thermal properties other than the insulative capacity of the building element. For example, thermal storage capacity becomes more important in order to delay or avoid internal heat radiation from the building element and to use the limited cooling effort only to reduce the internal air temperature.

Thermal storage and time lag properties

The amount of heat that is transferred through a material or building element is determined by the amount that is stored and the time it takes the heat to be transferred to its inner surface. The capacity that a material or element has to retain heat is referred to as its thermal storage capacity. Generally, the greater the capacity, the slower will be the transfer of heat gains made internally. The table below indicates the thermal storage capacity of various materials.

Approximate density and thermal storage capacity of various materials

Material	Density (kg/m^3)	Thermal storage capacity (W h/kg °C)
Solid burnt bricks	1800	0.26
Hollow bricks	1200	0.26
Stone	2600–2800	0.22–0.24

(cont'd...)

(...*cont'd*)		
Material	Density (kg/m³)	Thermal storage capacity (W h/kg °C)
Soil	1800	0.23
Timber:		
lightweight	400–500	0.55–0.65
heavyweight	600–800	0.55–0.65
Plywood	600	0.75
Grass fibre board	200–300	0.38
Compressed earth block	1700–1800	0.25
Clay (adobe)	1000–2000	0.23–0.30
River sand	1700–2000	0.22
Air (still air)	1.2	0.28
Water 10°C	1000	1.16
Concrete:		
light	1000–1700	0.5–0.9
dense	2400	1.1
Mineral wool and glass wool	20–120	0.17–0.19
Steel	7850	0.13
Aluminium	2700	0.26

(from various sources)

The delay in the transfer of heat is referred to as the time lag property of a material or a building element. The time lag will be greater with increase in the heat storage capacity. These properties provide an opportunity to store heat in a structure when the temperatures are at their highest and to release the heat in periods of low temperature.

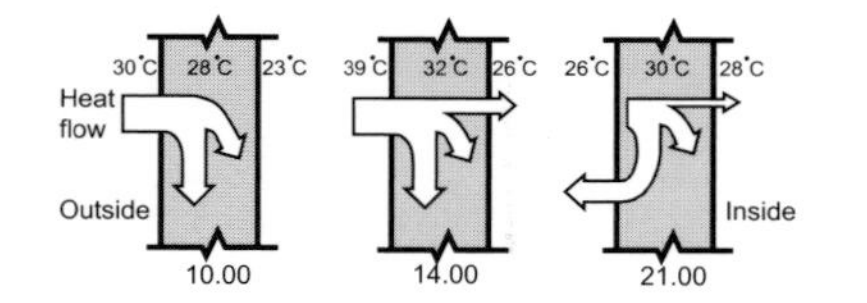

An example of heat flow in a wall with thermal storage capacity, showing outside, wall and inside temperatures as well as time of the day.

In composite building elements, the storage and the time lag capacity are complex to determine as the thermal properties of one layer will influence the amount of heat transferred to the adjacent layer. The heat-storing properties of one layer will determine the quantity of heat available to be transmitted to the next layer. The time lag properties of the building element will not be the sum of the time lags of the individual layers, but will depend on the combined properties and the sequence of the layers.

The time lag property is expressed in terms of hours. It is the time difference between the peak outer surface temperature and the peak inner surface temperature, or the time it takes the external heat to impact internally on a material or element. Below is a list of the time lag values of various materials and building elements based on a static situation with a fixed temperature difference between the outside and the inside. The values will be different in a real and dynamic situation, i.e. changes in outside temperature during the day and the dynamic influence of other building elements, etc.

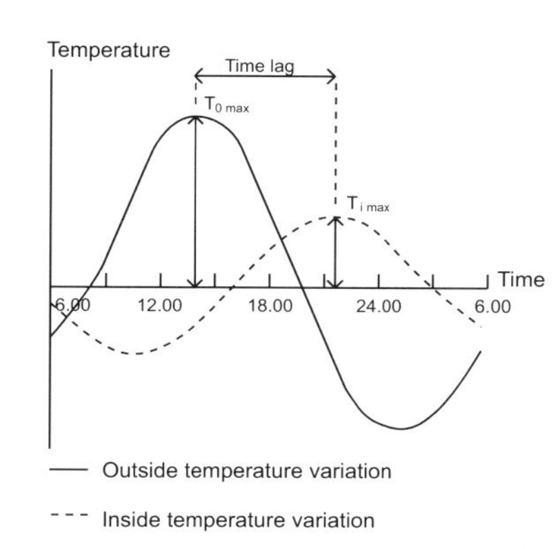

Definition of time lag.

Time lag for typical construction types

Construction type	Time lag (h)
Dense concrete:	
100 mm	2.5
250 mm	7.0
Brick:	
100 mm	3
200 mm	6
Wood:	
25 mm	0.5
100 mm	3.0
Stone:	
200 mm	5.5
400 mm	10.5
600 m	15.5
Stabilised soil:	
100 mm	2.5
200 mm	5.5
Unstabilised adobe, 400 mm	12
Stabilised adobe, 250 mm	9
Rammed or compressed earth, 250 mm	10
Cavity double brick walls	6

(cont'd...)

(...cont'd)

Construction type	Time lag (h)
Cavity wall with massive double concrete block (2 x 115 mm), with 15 mm cement render	10
As above, but with hollow concrete blocks	11
Cavity double wall of 100 mm hollow terracotta blocks, both with cement render	9.0
100 mm reinforced concrete slab with 40 mm glass wool insulation under the slab	3.0
As above, but with insulation on top of the concrete slab	12

(from various sources)

Cavity walls and roofs

Composite walls have many advantages, especially in hot dry zones. Their efficiency depends primarily on the surfaces of the materials enclosing the air cavity. If their surfaces are reflective, heat transfer between them is considerably reduced. This reflected heat could be removed through air vents placed high in the walls, bringing about an inflow of cool air through low vents.

A method of keeping the ceiling cooler is to use a composite roof construction. In warm humid zones it can be constructed from two lightweight layers, and in hot dry zones an outer light layer and a massive inner layer can be used. In both instances the main function of the outer layer is to shade the inner layer and to minimise its heat absorption by reflecting as much solar radiation as possible. The efficiency of the outer layer will also depend on its absorptivity and emissivity properties. The inner layer should be either heat storing or heat insulating. To reduce radiant heat in the enclosed cavity, a highly reflective surface can be used on the inner layer. The prime requirement when using composite roof construction is the removal of accumulated heat through adequate ventilation of the cavity.

Design principles: thermal properties of roofs, walls and floors

Hot dry zones

- External surfaces with good reflectivity properties will reduce heat gains during the day.
- External surfaces with good emissivity properties will allow a greater amount of the heat gained during the day to be released by the structure at night.
- Dense materials for roofs, walls and floors will even out diurnal temperature variations.
- Walls will experience reflected radiation from the ground and adjacent building surfaces. Wall surfaces should have good reflective properties and be light in colour.
- Lightweight materials with good insulative properties can be used in combination with thermal storage materials to create composite walls or roofs.
- In composite roofs and walls lightweight materials placed near the outside surface and dense materials near the inside will generally create a greater time lag than if placed the other way round.
- The time lag of roofs and walls must be planned according to the function and time of use of rooms directly adjacent to them. For example, offices and classrooms, which are used during the day, may require only a 6–8 hour time lag, whereas rooms that are used throughout the whole day may require time lags of between 9 and 12 hours. A room used only for sleeping could have a very short time lag in order to profit from the night-time cooling effect.
- Air gaps between building elements causing infiltration of air should be eliminated. Buildings in hot dry regions require that the entry of heat/air be controlled.
- Concrete floors, terrazzo tiles and other hard, cold materials are comfortable internally in a building. These materials will not feel hot on bare feet even with high surface temperatures.

- Floors should maintain direct contact with the ground to take advantage of the ground's heat-storing properties.

Warm humid zones

- The use of lightweight and low heat-storing materials is preferable to ensure minimal radiation towards the inside.
- Reflective outer roof surfaces can be used to reduce heat gains. Reflective materials, e.g. shiny metallic surfaces, should not be used for walls as these will create uncomfortable outdoor environments due to reflected glare.
- Insulating materials can be used for roofs and walls as long as care is taken not to hinder possible night cooling and to ensure that heat is not trapped inside the building. Insulating materials should not create condensation problems from temperature differences on the surfaces of the building element. Condensation can also produce humidity in the insulation material itself, which will reduce the insulation effect considerably.
- Materials for roofs and walls should be selected so that they do not allow internal heat gains or create indoor air temperatures that are higher than outdoor air temperatures. Heat-storing (dense) materials should be avoided unless the intensities of solar radiation impacting on the surfaces are very high, and if they cannot be readily shaded as on east- and west-facing walls. If heat storage is required in order to compensate for diurnal temperature ranges, dense materials can be used for internal walls.
- Floor materials should have low thermal storage properties.
- Floors should be elevated above the ground as a means of promoting air movement from below and reducing heat gains from the ground.

Design principles: thermal properties of openings

The greatest source of heat gain occurs through openings. Glass is especially transparent to short-wave radiation emitted by the sun but almost opaque to long-wave (heat) radiation emitted by internal objects. Once solar radiation has been transmitted through a glazed opening, it is trapped inside the building (the greenhouse effect).

When solar radiation is the major source of heat gains, there are four methods that can be used in its reduction:

- orientation (see Chapter 3);
- external shading devices (see Chapter 5);
- internal blinds and curtains;
- special glass.

Orientation and external shading devices have been discussed in detail in the sections mentioned above. The two remaining variables are discussed below.

Internal blinds and curtains

Internal blinds and curtains are not an effective means of solar control. They will stop the passage of radiation, but they themselves absorb heat and can therefore reach very high temperatures.

The absorbed heat will be partly convected to the indoor air and partly re-radiated. Half of this radiation is outwards but, as it is long-wave, the glass will stop this process. Consequently, the space between the opening and the blind or curtain will be substantially overheated.

Special glass

On opaque surfaces radiation is partly absorbed and partly reflected. With transparent surfaces, it may be absorbed, reflected and transmitted. A range of glass is presented below that may improve the thermal performance and the quantity of light. However, the quality of the light, i.e. colour and intensity, will be influenced, which is also a consideration.

Ordinary window glass transmits a large proportion of all radiation. Varying the composition of the glass can, however, produce selective transmittance of the glass, which affects heat and light transmission.

The are two main types of selective transmission glass: heat-absorbing and heat-reflecting glasses. These types of glass can be combined in two or three layers in a thermal glass construction with air or gas in the space between the glasses. Reflective glasses are normally coated, and are available in a range of reflectances and colours.

Below is a list of some types of glasses indicating their light transmission and heat transmission values as well as their insulative properties.

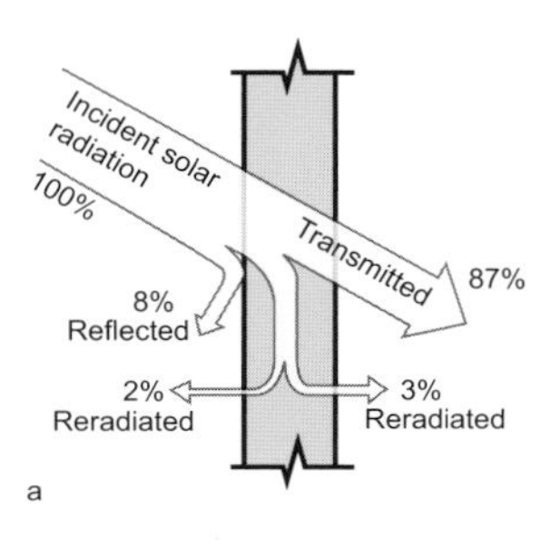

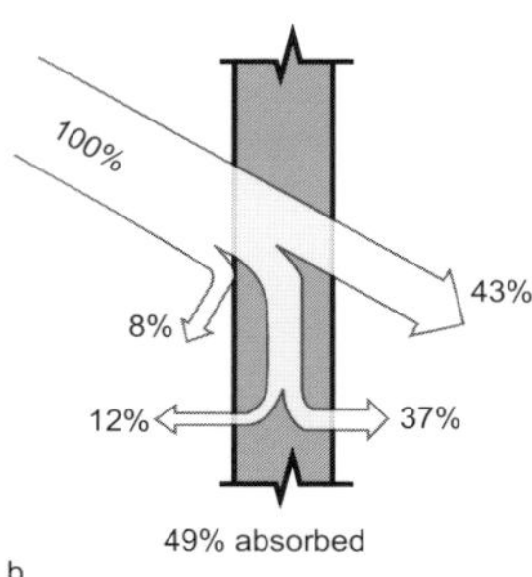

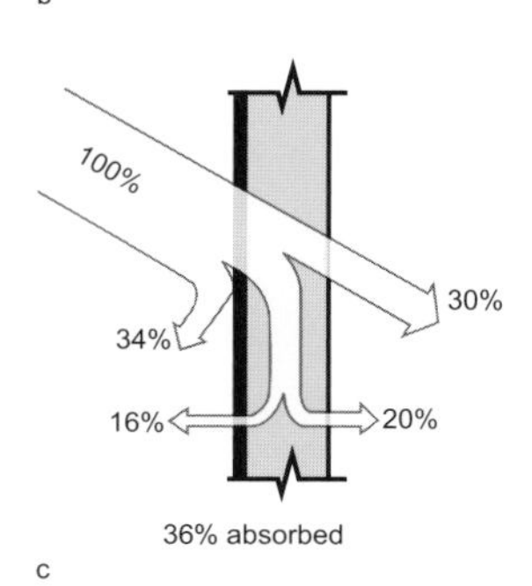

Examples of heat transmittance through: a. ordinary glass, b. absorptive glass and c. reflective glass.

Examples of transmission properties and U-values of various types of glass

Type of glass	Light transmission value (%)	Heat transmission value (%)	U-value (W/m² °C)
Ordinary 6 mm plate	87	90	5.6
6 mm heat absorbing	43	80	5.6
6 mm heat reflecting	30	50	5.6
Double glazing: 4 mm + 12 mm gas + 4 mm, ordinary glasses	77–85	59–84	1.3–2.7
Double glazing: 4 mm + 12 mm gas + 4 mm, reflective glass outside and ordinary inside	56–68	35–46	1.2–1.5

(from various sources)

Translucent thermal insulation

Translucent thermal insulation with high insulative and light transmission properties is currently being developed. This type of wall or ceiling element could in the future play an important role in maintaining an appropriate level of daylight through openings that will not generate heat gains in buildings in hot climates. The glass brick was one of the first kinds of translucent thermal insulation. It has approximately twice the insulation capacity of ordinary glass. The glass brick has, however, rather low light transmission values.

7. Natural ventilation and cooling

Natural ventilation may supply fresh air and aid the removal of odours and the removal of internal heat, cool the structure and reduce structural radiation, as well as produce evaporative cooling of the body and the air.

Natural ventilation can occur due to:

- *wind-generated air pressure differences;*
- *temperature-generated air pressure differences.*

Natural ventilation in the hot dry zone is a mixture of vertical and horizontal air movements driven by either temperature- or wind-generated pressure differences, whereas in warm humid climates ventilation is normally horizontal and based on wind-generated pressure differences. Natural ventilation based on horizontal air movements is not suitable for sites with external noise or air pollution (urban environments). Stimulated ventilation by simple and energy-efficient mechanical equipment is also considered to be a natural (passive) ventilation method.

Natural cooling can produce cooled air or radiative cooling, i.e. the internal building surfaces are cooler than the ambient air or the body. A selection of natural methods is presented below.

The functions and requirements for ventilation

Ventilation needs in hot dry and warm humid climates based on passive measures should not be confused with standard requirements for odour removal and fresh air intake, which are stipulated by building regulations and advisory codes in temperate climatic zones in industrialised countries.

The fresh air intakes in codes and standards are normally expressed in terms of the required number of air changes per hour and/or as a number of cubic metres of fresh air per hour per person for different spaces and activities. They serve primarily as performance requirements for mechanical ventilation or air-conditioning systems where rigid rules can be applied.

Natural ventilation in hot dry and warm humid climates embraces several different functions, which vary in accordance with the time of day and night and the time of the year. An air speed that is much higher than for temperate climates is needed, especially if evaporative cooling of the body is included. On the other hand, air changes may also need to be reduced to less than the minimum requirements for odour removal, for example during extremely hot days in the hot dry zone, because the internal hot air will be replaced by even hotter external air if cooling of the incoming air is not achieved, or is impossible. Night-time ventilation is useful as a mean of structural cooling in periods with high daytime temperatures and considerably lower night-time temperatures. For structural cooling the rate of change of the air should be considerably higher than that necessary for odour removal, but less than that for cooling the body.

Below are two tables from *Housing Climate and Comfort* by Martin Evans. The first table introduces ventilation functions and their typical requirements.

Functions:	Fresh air and odour removal	Structural cooling	Cooling the body
Requirements:	1 air change per hour	10 air changes per hour	1–2 m/s (equivalent to about 100 air changes per hour)

It should be noted that wind speed and air change rates are not related. For example, high air speeds can be found in a room close to a ceiling fan even though the air change rate is very low, and an open building exposed to winds might experience low wind speeds at the level of the occupants, but a relatively high air change rate. The following table shows examples of ventilation requirements primarily for odour removal. Uncontrolled ventilation/infiltration of air from air gaps in badly constructed or maintained buildings may in many cases amount to more than the minimum ventilation rate (1 air change per hour) for the whole building.

Examples of common ventilation requirements for fresh air and odour removal

Space and activity		Ventilation rate (air changes per hour)
Space for living or light office work:		
Occupancy density:	5 m^2/person	1.2–2.0
	10 m^2/person	0.4–0.7
	15 m^2/person	0.1–0.25
WC (4.5 m^3)		3
WC and bathroom (12 m^3)		1.5
Kitchen (approx. 10 m^2) to prevent condensation:		
Kitchen with non absorbent surfaces:		
gas cooking		13
electric cooking		9
Kitchen with absorbent surfaces:		
gas cooking		5.5
electric cooking		2.7
Minimum for whole building:		
Minimum		1
To avoid odours and stuffiness		2
To avoid condensation		4

Notes

The book *Healthy Buildings* by Bill Holdsworth and Antony F. Sealey has, along with other books, interesting discussions on internal air quality, material emissions and other sick-building syndromes that affect the ventilation needs. It is interesting to note that about 42% of the perceived pollution in a modern office building comes from the air-conditioning system and 20% from materials emissions (according to the report *Less is More* produced by the Thermie Programme Action by the European Commission Directorate-General for Energy, quoting a study by P. O. Fanger). The remaining perceived air pollution comes from humans, i.e. smoking and the body.

Wind-generated air pressure differences

Ventilation in a building based on wind-generated air pressure differences can potentially create higher wind speeds or air changes rates than required and is anyway more difficult to manage than temperature-generated air pressure differences. The directions and speed of winds can, however, be adapted to maximise their potential for natural ventilation. The siting, orientation and layout of a buildings and its surrounding features, as well as a building's form, can be used to manipulate the wind before it enters a building.

In the same way that winds are generated by pressure differences in the atmosphere, air movement through a building is the result of air pressure differences created between two sides or two openings. When air strikes an obstacle such as a building, it will exert pressure on the obstructing surface. Vortexes will also occur whenever a building obstructs and separates the flow of air. On the windward side of a building such vortices are at an increased pressure and on the leeward side they are at a reduced pressure. If a building has an opening facing a high-pressure zone and another facing a low-pressure zone, then air movement will be generated through the building.

Siting, orientation and layout

Siting and wind orientation have also been discussed in Chapter 3, The built environment. In urban planning the urban form, layout and spacing can be used to achieve the desired air movement. The

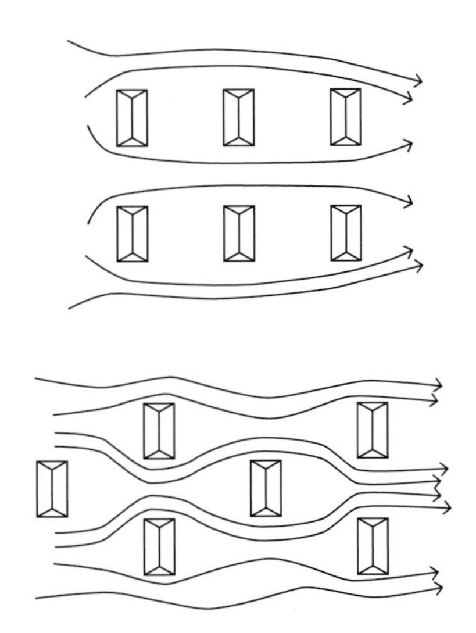

Urban layout and air movement: A grid layout creates wind-sheltered zones. A staggered layout will allow for a more uniform air movement, with buildings being exposed to winds, and have positive air pressure on the wind side and negative pressure on the leeway side to facilitate ventilation.

siting of groups of buildings and their effects on air movement will be one of the most important considerations. New buildings can be sited so that air movement patterns are altered or enhanced to improve the ventilation conditions for more buildings.

When buildings are laid out in a grid, spacing of six times a building's height will generally ensure air movement for the buildings, but there might be a suction effect on all facades, as well as on those perpendicular to the direction of the wind, which will block air movements. In a similar situation, but with the buildings staggered, the air flow will, however, be more uniform to assure pressure differences which are necessary to achieve cross-ventilation.

The relationship of the elevations of the buildings, i.e. their height and spacing relationships, will also have an influence on air movement.

The form of the individual buildings will create areas that are wind sheltered, or where wind is deflected. This will have an effect on air movement patterns surrounding a building and on the comfort conditions of outdoor spaces.

The deeper and wider a building, the smaller is the wind shelter behind it. Whereas the wind sheltered area of a building is directly proportional to its height.

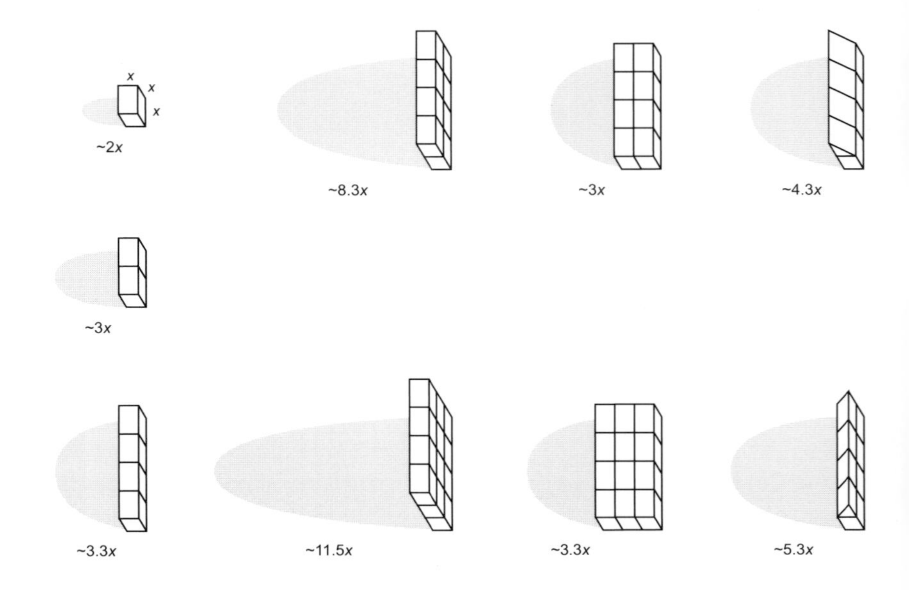

The deeper and wider a building, the smaller is the wind-sheltered area behind it. The extent of a building's wind-sheltered area increases proportionally to its height. The pitch of a building's roof will also have an effect on the sheltered area/air movements. (from various sources).

Surrounding features

Landscaping can play a vital role in redirecting air movement (see also Chapters 3–5). Air movement should be directed to pass through shaded areas and not across heated surfaces. In hot dry climates, vegetation can be used to stabilise dust movement and reduce dust storms. This will allow air to be both cooled and filtered before entering a building.

Windbreaks can enhance air pressure differences around buildings. External structures, such as walls adjacent to buildings, will strongly influence the pressure built up on a building's facades. These walls can be created from dense planting such as hedges, or can be built structures. They can be used to alter the direction of winds, deflect undesired winds or change the velocity of winds. For example, wing walls located at the downwind end of a building increase the positive pressure build up on a facade considerably. The opposite, a reduction in pressure, will occur if these projections are placed upwind from openings. Hedging will allow a gentle breeze to filter through.

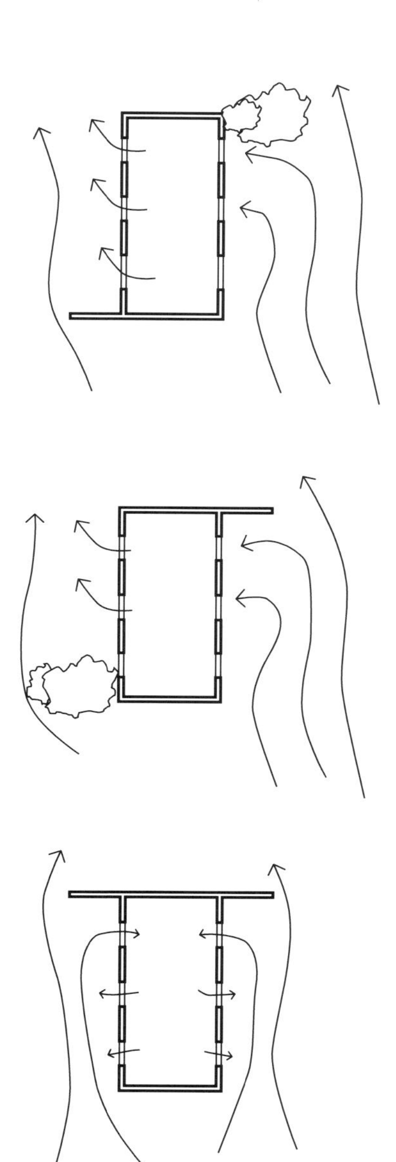

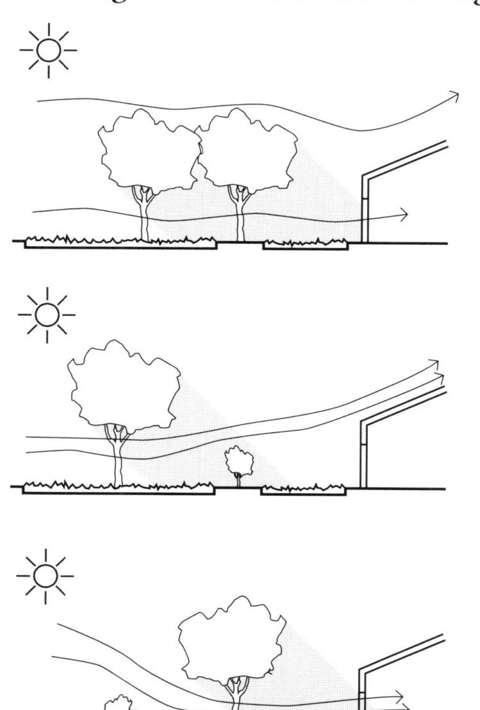

Walls and planting can be used to direct air movement.

(left) Airflow patterns will be influenced and can be modified by landscaping. Hot air can be cooled by passing over or through vegetation before entering a building.

(right) In warm humid regions the structure should be open, light and air transparent. The maximum distance between the opposing facades is normally set to 15 m. Double-banked rooms will limit natural ventilation unless other means of stimulating the air flow are used. The maximum depth of a not-cross-ventilated room is normally set to 6 m.

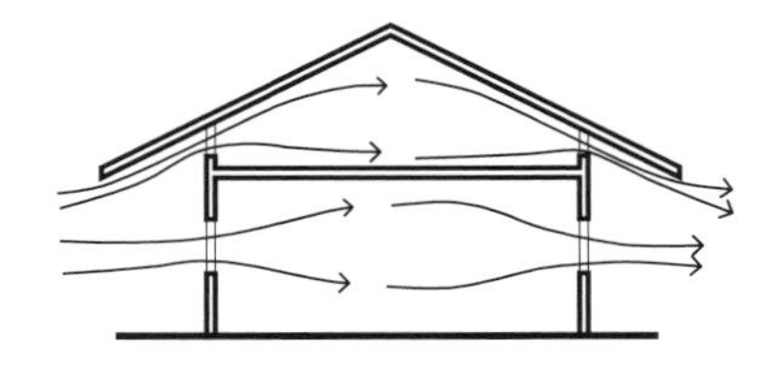

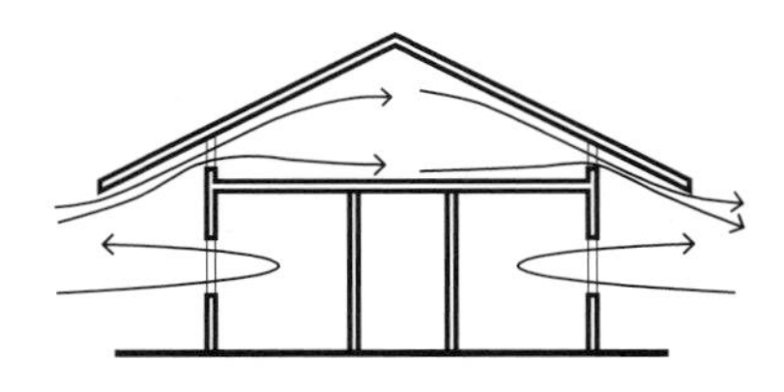

The building envelope and its internal layout

The following is a summary of the ventilation principles that can be adopted when making design decisions regarding the building envelope's form, shape, projections and its relationship to adjacent buildings. The effect of the building plan and its internal layout is also discussed.

In hot dry climates air movement from the outside can, during the midday hours, increase discomfort as the air is very hot. Under these conditions it is necessary to reduce ventilation and exclude external air movement unless the incoming air is cooled before entering. Ventilation is required at night to cool the interior of the building and its occupants. One method of achieving this is to form the building around a shaded and planted courtyard. Air can be drawn from the courtyard into the building through low openings. (See Chapter 4, Special topic: courtyard design.)

The compactness of a building, e.g. with double-banked rooms, has the advantage of reducing solar gains due to low surface area to volume ratio. However, the depth and the internal walls of the building will affect, and can eliminate, the effect of natural ventilation. A cross-ventilated room may have a depth of 10–15 m depending on the effectiveness of the ventilation opening (size, form and number) whereas a non-cross-ventilated room should not have a depth of more than 6 m.

In warm humid climates, where cross-ventilation and good internal air flow are required, buildings should be narrow (e.g. single-banked rooms) and with openings on the two opposing facades to promote cross-ventilation.

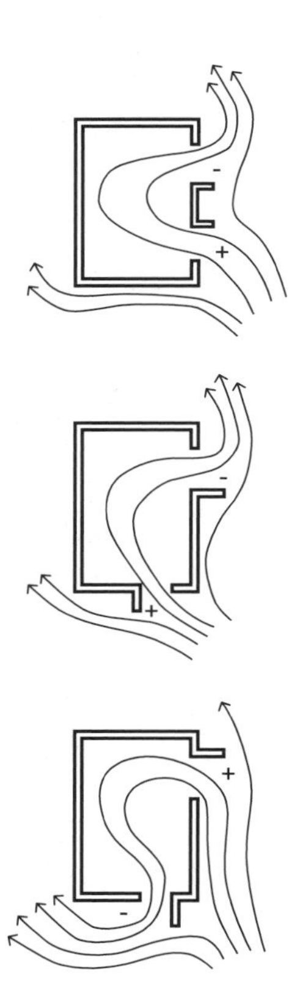

Illustrations from studies by B. Givoni on natural (horizontal) air flow in a building or room with two openings with different position and different types of wing walls.

B. Givoni has made extensive studies of the effect of projections and has found that the ventilation potential can be increased considerably. External projections that may also serve as shading devices can be placed vertically and horizontally to improve the

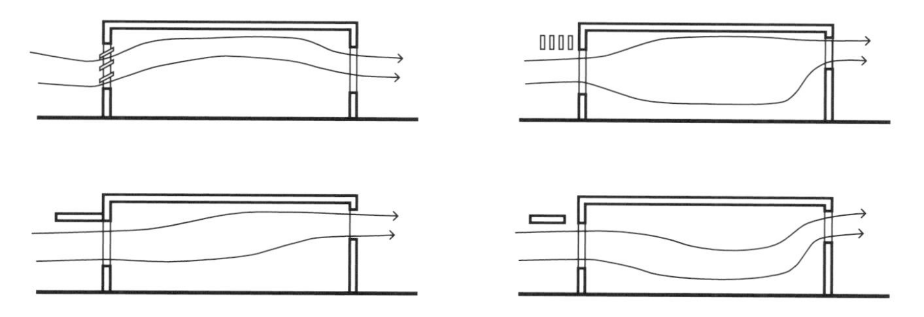

Air flow through openings in vertical section. The effect of openings fitted with louvres or shading devices on internal airflow.

air flow. A breeze can also be directed towards the objects of ventilation/cooling, i.e. cooling ponds, structures or occupants.

Other means of stimulating the air flow are the use of the stack effect or of simple mechanical stimulation, discussed below.

Openings for ventilation openings and openings for light are not necessarily the same openings (see Chapter 5, Opening design). The incoming air could enter from specific ventilation-only openings in the walls or pass through a cooler space such as an underground duct, the subfloor or a shady exterior. These solutions are appropriate for both hot dry and warm humid climates.

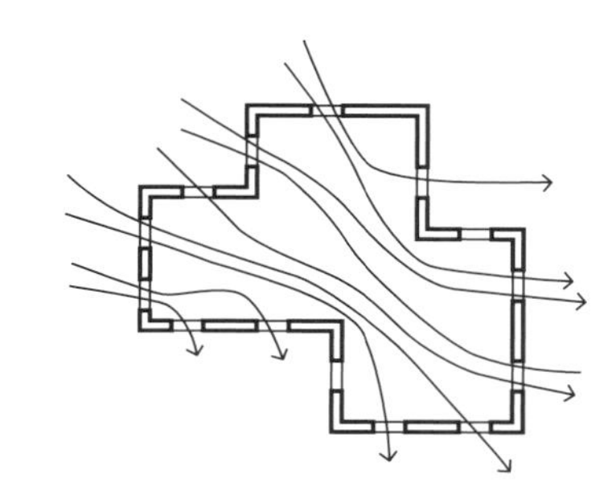

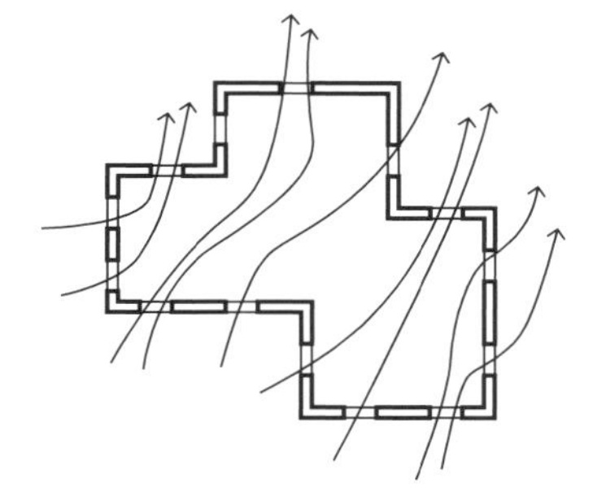

Internal layout

Air movement can be impeded by inappropriately designed internal partitions. When a wall divides a room, or if several rooms occur together, and doors or corridors separate air inlet and outlet openings, air will naturally change direction and speed. Satisfactory ventilation is possible in buildings when air has to pass from one room to another, as long as the connection between the spaces is of appropriate size and remains open when ventilation is required. However, the quality of ventilation in residential or working areas does not depend only on average air speeds or air changes. The variation of these parameters in the space must be evaluated, and stagnant pockets of air without air movement or air changes should be avoided as much as possible.

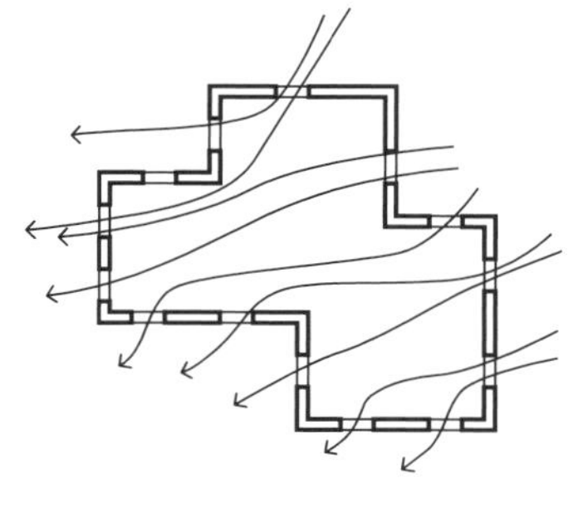

Illustrations from studies by B. Givoni of a building plan and its capacity to adapt to different direction of winds for natural ventilation.

Air velocity is at its lowest when partitions are placed close to the inlet opening, as the air is forced to change direction rapidly. Air flow will also be reduced in speed each time it is forced around *a&b: show the effect of inlet and outlet opening sizes on natural ventilation. a: When the air inlet is larger than the outlet the air velocity is diminished in the room and is increased just outside the room as the air is forced to go through a smaller area. b: The opposite happens here, and the ventilation rate/air changes in the room will be considerably higher but the ventilated area is smaller than in a. c&d: show the effect of internal wall on airflow and distribution of natural ventilation.*

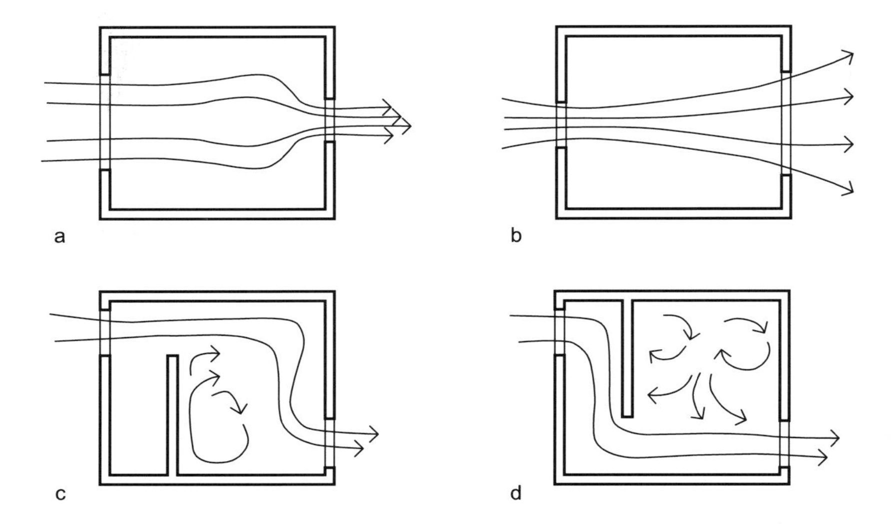

or over obstacles. Where partitions or walls are unavoidable, some air flow can be ensured if partitions are lifted off the floor and finished below the ceiling.

Air velocity is at it highest when it is forced through constricted areas. For certain wind speeds this may create (unwanted) suction/ventilation effects in adjacent rooms.

Temperature-generated air pressure differences

If a building has two openings at different levels in two zones, e.g. two opening inside or one outside and one inside, and if there is a temperature difference between the two zones, the stack effect operates. It is the natural tendency of warm air to rise and escape through the top opening. The rising hot air will be replaced by cool air entering through low openings. The simple effect of cold dense air moving down and hot less dense air rising is called the buoyancy effect, which will assist the stack effect as a cooling agent.

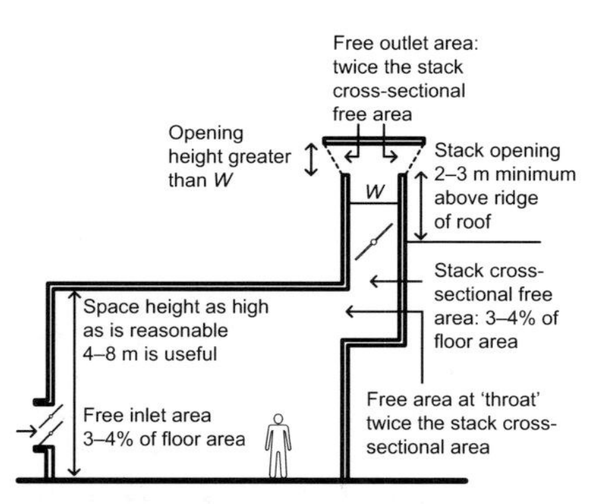

Rules of thumb and design aspects for stack-driven ventilation. From Environmental Design, *edited by Randall Thomas.*

The air velocity achieved by the stack effect alone is usually not enough to provide a cooling effect for our bodies directly, but it is useful as a way of expelling the build-up of warm air inside a building and for night-time structural cooling, so having a cooling effect by letting in cooler air. In areas or at times where the wind speed is low or non-existent, the stack effect can be used to generate internal ventilation.

The size and shape of the openings and the distance between them, can be used to control the ventilation, i.e. air volume and speed. In addition, elements such as wind towers or extract fans can be used when outside wind pressure differences are available in order to increase air movements and direct air flows.

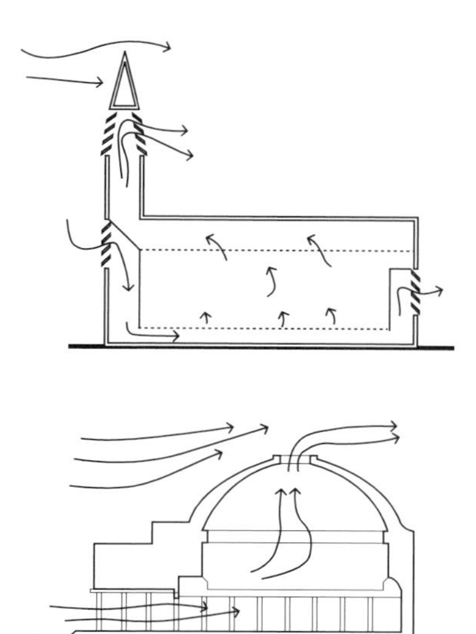

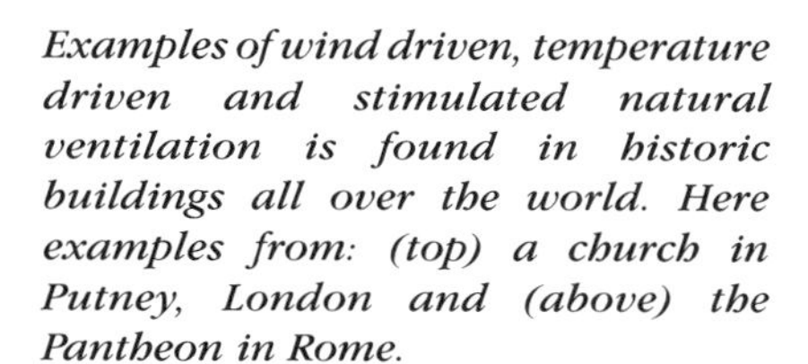

Examples of wind driven, temperature driven and stimulated natural ventilation is found in historic buildings all over the world. Here examples from: (top) a church in Putney, London and (above) the Pantheon in Rome.

STIMULATED NATURAL VENTILATION

Improvements of natural ventilation due to the form, shape, projections, etc. of the envelope of the building are mentioned above. A collection of component and designs that are based on air pressure differences and/or the stack effect to stimulate the natural ventilation further are discussed below.

Wind towers

Wind towers have been used for centuries in the hot dry climates of the Middle East. Traditionally these natural ventilation towers rely on external winds to cool the interior of a building, and they were commonly supplemented with devices to cool the incoming air. The air that enters the wind tower will, in the hot dry climate of the Middle East, have a low moisture content. If water-filled, unglazed earthen pots or a water-soaked cloth or wet charcoal are placed in the air inlet, the incoming air is cooled as the water evaporates.

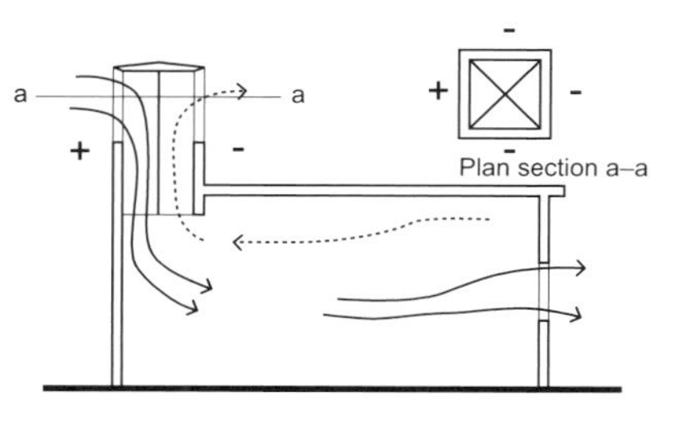

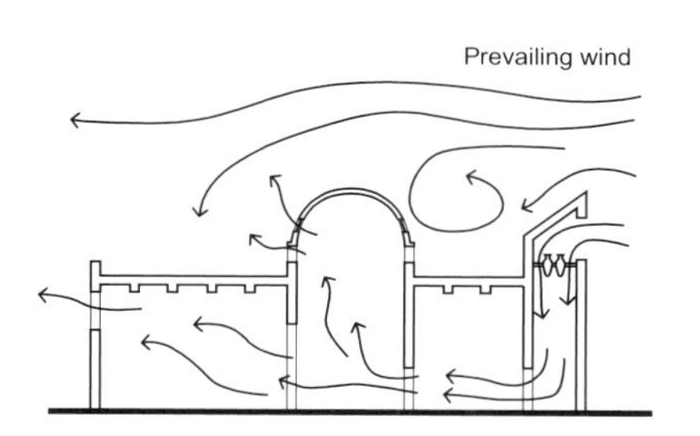

(left) A traditional multidirectional wind tower where wind scooping occurs at the windward side(s) and wind suction on the leeward side(s). (Right) A traditional Middle Eastern wind tower for wind scoop where wind passes across porous clay jars filled with water. Hot air is released through high openings.

The inlet openings of wind towers are normally designed to reach up to the relatively uninterrupted, cooler and less dusty wind flow in order to catch the wind and scoop air down into the buildings. Outlet openings are placed on the roof and/or high up on walls on the leeward side of the building structure, where the pressure is less than inside the building and a suction effect occurs.

The design of the wind tower depends upon the wind characteristics of the area. If wind directions are relatively predictable, it can be designed to be unidirectional. If, however, wind directions are inconsistent, it is designed to be multidirectional. The traditional multidirectional solutions were normally designed to have two or four vertical ducts, which give the added advantage that suction will occur on the leeward side of the tower, making the tower capable of scooping air into the building on one side and at the same time acting as an outlet/suction system on the other side(s). Studies of the available literature on buildings with natural ventilation indicate inlet and outlet opening areas for wind towers to be 3–5% of the floor area they serve.

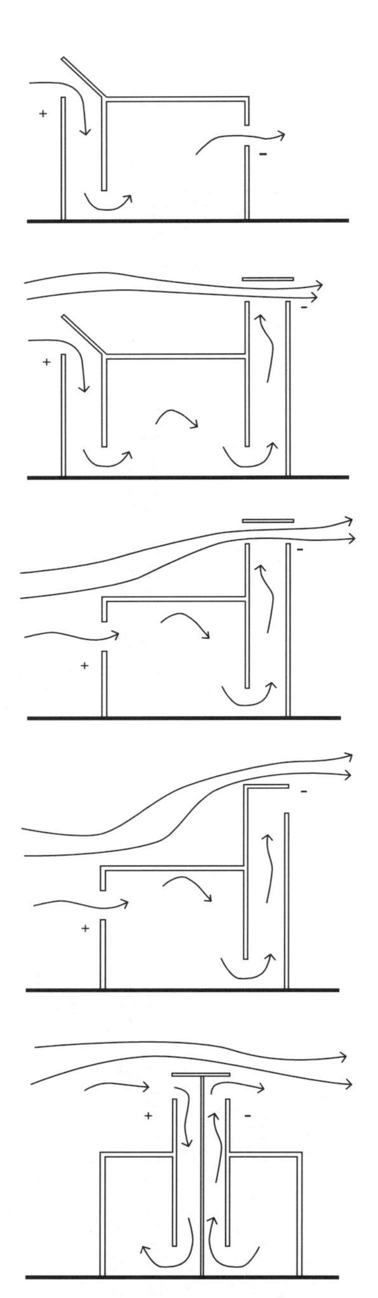

Wind tower design principles.

(right) Simple wind tower from Pakistan. They are constructed from mud, canvas or matting.

When there is no wind, any type of wind tower will act as a ventilation (suction) unit as the process of air movement is the reversal of the scoop effect. This is because the outside air pressure is lower than that of the air in the building, with the result that warm internal air can move out over the top of the tower, driven by the stack effect. The stack effect alone, however, is less efficient than the wind effect. Total wind calm is rare, and the two effects are normally combined. Modern designs of natural ventilation towers utilise wind or the stack effect or both in a similar way to the traditional methods. They are often equipped with mechanical fans to enhance the movement of air in order to be flexible enough to address changing wind and temperature conditions. Modern designs of wind towers, roof and building forms as well as roof devices for naturally ventilated buildings are today designed more aerodynamically than the traditional structures, i.e. with shaped and rounded forms to enhance the air flow.

A multidirectional wind tower from Yazd, Iran. (below) Wind tower for scooping air from Sindara, Pakistan. (below left) Multidirectional wind towers from Qatar University, Doha. Dr Kamel El Kafrafi, Engineers Ove Arup & Partners.

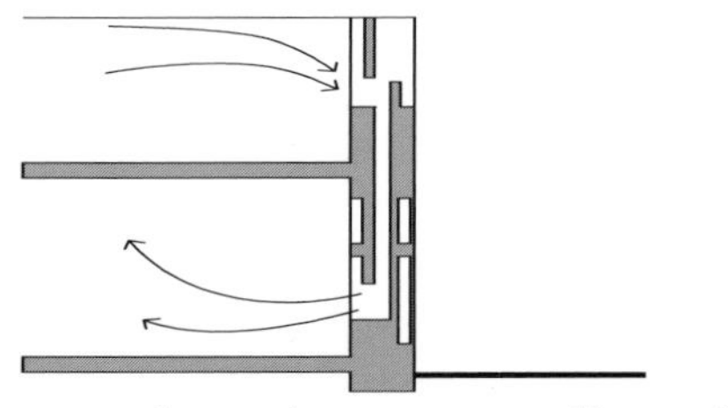

A traditional parapet wall windcatcher.

(below) An example of a modern energy-efficient double-skin facade system. Based on drawing from the ALCO Twinface facade system, Germany.

(below right) A commercial building in Harare, Zimbabwe, with structural cooling and natural ventilation. The illustration shows the overall airflow in the building. Architects: Pearce Partnership. Engineers: Ove Arup & Partners.

Ventilation of wall and floor structures

Structurally integrated windcatchers, air ducts or vents can be incorporated into the design of walls and have traditionally been used in the hot dry climatic zone. For example, recessed niches on the external wall, at the floor level and in the roof parapet can be used to create slots between two vertical structural posts. This will reduce the temperature of the incoming air as heat is absorbed by the structure over which the incoming air is passing.

When wall and floors are ventilated they not only can absorb the heat from the outside, but can also be used to absorb and store the heat load created internally in the building by occupants and equipment. The stored heat can be ventilated out of the building by convection and radiation to the ventilation system in the floors and walls. A ventilated wall may also have a temperature lower than the air temperature and act as a radiation cooling agent.

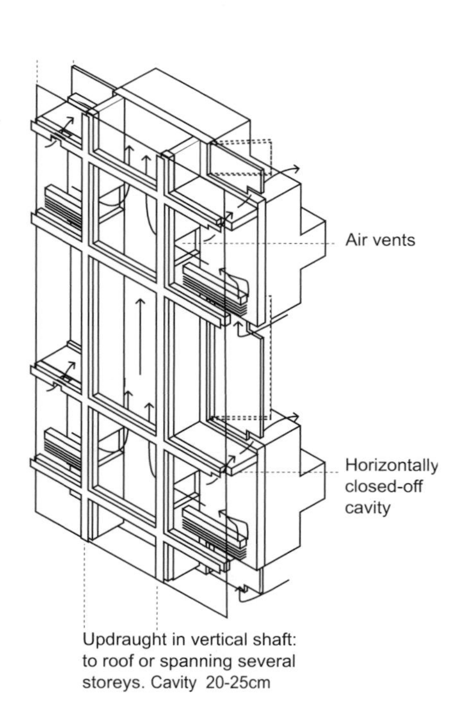

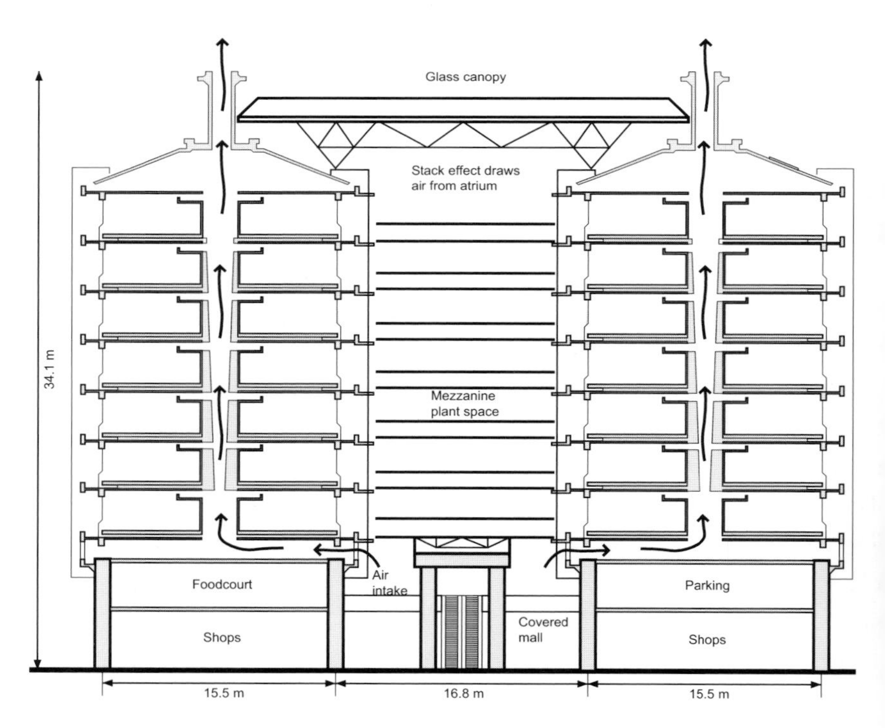

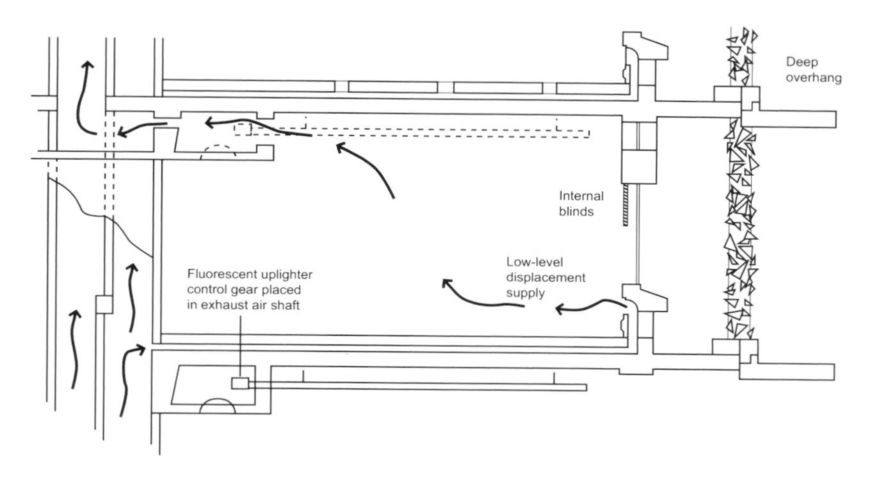

Same building as below right on the previous page. The illustration shows details of airflow in office spaces.

Double skins, integrated or intelligent facades are the names of new forms of facade walls where, for example, ventilation, solar control and insulation are combined. These facade types have, amongst others, the advantage of limiting the nuisance of noise in urban environments, and are often designed to operate more or less automatically intelligently. They have primarily been used and developed for climatic regions other than those in this Guide.

Roof ventilation devices

Roof ventilation devices can be incorporated into a roof to draw hot air out of or cool air into a roof cavity or a building's interior. They can be designed to utilise external winds for direct scoop or suction to enhance the stack effect. A combination can be used to cater for changing climatic conditions.

These devices can be unidirectional, orientated to capture the most favourable winds for scooping, or multidirectional or rotational devices that can capture winds from various directions. The same design options can be used for the suction effect. The shape of the roof devices and the surrounding building elements are important when evaluating the effectiveness of a specific system.

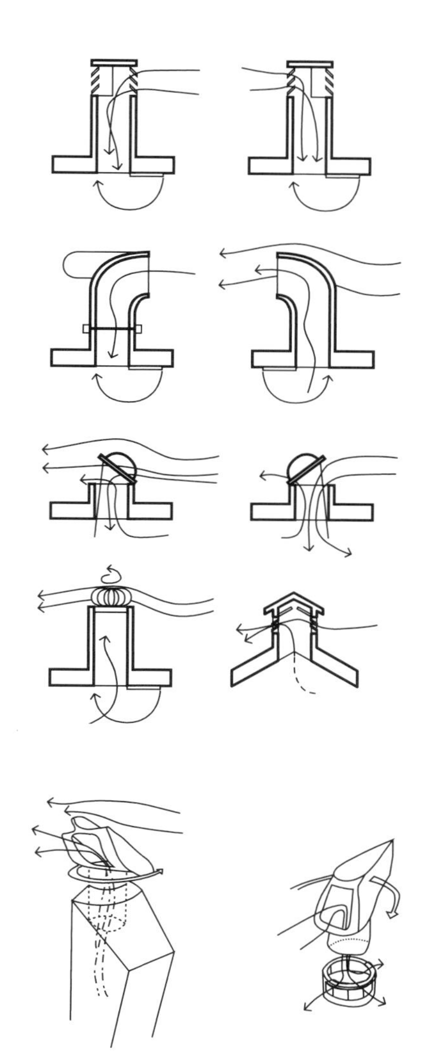

Examples of various simple roof ventilation devices, as well as sophisticated examples of aerodynamic design for wind suction and wind scooping.

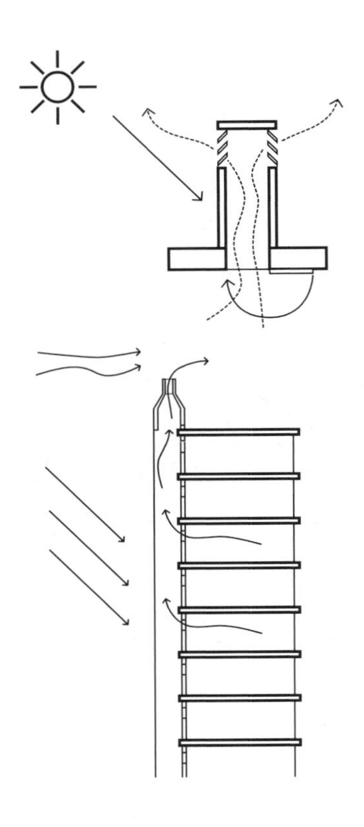

Schematic examples of a solar chimney: as a roof device painted black and as a facade installation with glass towards the sun to capture the heat. Both measures will enhance the stack effect, in particular the facade installation.

(right) A combined roof ventilation device. It possibly combines the solar chimney, wind tower and stack effect and may also function as a skylight. Found on a house in Darwin, Australia.

Solar chimneys use the sun's energy to induce the stack effect and to stimulate air movements under still conditions. The chimney is equipped with a heating element that creates a natural draught as the heated air rises. In its simplest form, it is a chimney covered on the outside with black metal, which absorbs heat. As the temperature increases in the chimney, the stack effect will draw the interior air up and out of the chimney. As the day becomes hotter the solar chimney improves in efficiency. The use of glass and/or materials with high thermal storage properties can further increase its effectiveness so that the chimney continues to function on stored heat even after the sun has set.

Notes: The venturi effect

The roof ventilation device shown in the margin to the left is supposed to create the venturi effect. Air is encouraged to move through a constricted area whereby its velocity increases: this produces a pressure decrease and a suction effect in the tower. The design is not, however, recommended unless the opening between the cover and the top of the tower can be automatically adjusted. The opening between the top cover and the tower will create a venturi effect only within certain limits of wind speeds, and will have no effect if the wind speed decreases and the opening size is maintained. At higher wind speeds the wind will

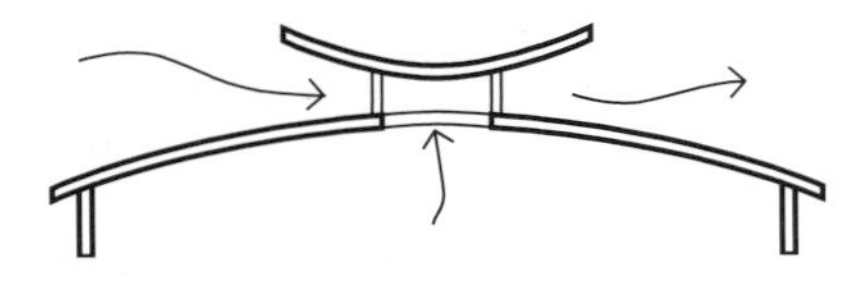

The venturi effect.

pass over or around the tower rather than across the opening. 'The net result is a system less effective than a simple open-piped chimney' – a quotation from *Wind Towers* by Battle McCarthy Consulting Engineers.

A master in chimney design was the architect Antonio Gaudi from Spain. His designs might inspire future design of solar chimneys, wind towers or roof devices. They are not just only sculptured! The chimneys on the photo are from Casa Milà in Barcelona.

Forced ventilation

Forced ventilation or air movements should be used to compensate for climatic impacts rather than poor building design. Where natural wind speeds are low or buildings have to be closed owing to extreme external conditions mechanical aids are the only possible means of inducing internal air movement.

Simple mechanical devices

The advantage of using fans to stimulate air movement is that the effect is more controllable and can be directed as needed (e.g. away from work areas where papers or other material can be disturbed). They can be small, simple and designed to operate at times when external wind forces are low, and can be located in positions where controlled air movement patterns are critical, e.g. in roof cavities.

Personal fans can be used close to occupants and directed in such a manner that others are unaffected. Larger ceiling-mounted fans provide a broad area of air movement and rotate slowly, creating gentle air movement without noise. This type of fan is ideal in cases where air movement is needed, e.g. for cooling the body in a non-ventilated space. An ordinary ceiling fan produces approximately 50 W of heat, and care should be taken that the number of fans are correctly calculated. More energy-efficient fans and less heat-producing ceiling fans have recently been developed.

Air extract and intake fans can be used for situations where the natural ventilation system is not adequate for the ventilation requirements. These can be simple devices designed to operate on solar energy. Options for mixed systems with air extract fans and air supply via wind scoop or air intake fans and extraction via a wind tower can also be applied.

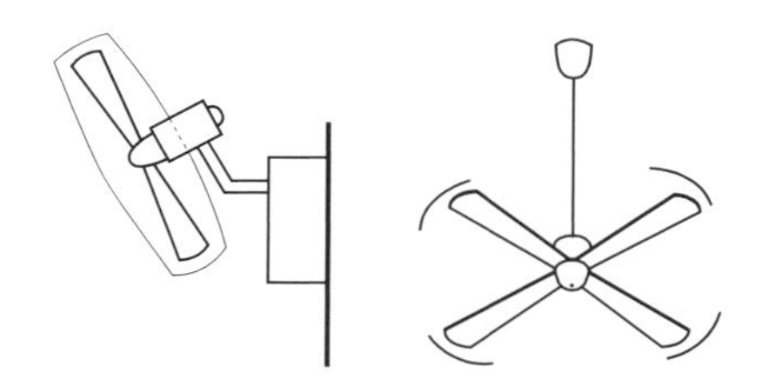

Examples of a wall-mounted and a ceiling-mounted fan.

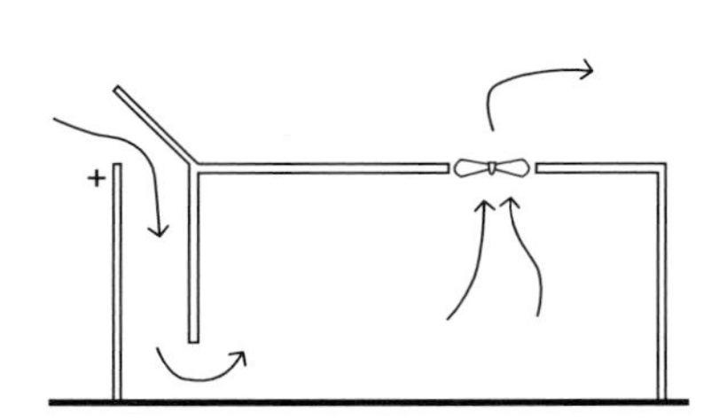

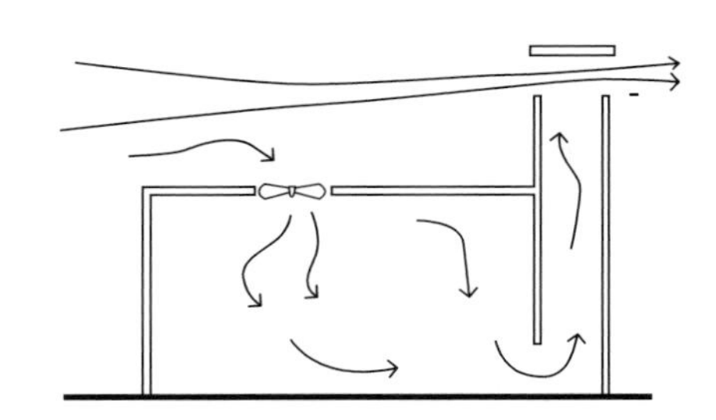

Schematic examples showing the use of simple mixed ventilation systems.

NATURAL COOLING METHODS

Varying degrees of cooling can be achieved in a building using natural sources such as the following.

- Evaporative cooling: the evaporation of water to cool the air and/or the building structure. (The effect of evaporative cooling of the body is described in the section in Chapter 2, Thermal comfort requirements.)
- Earth cooling: Use the relative cold subsurface soil to cool the building external surfaces or to cool the air in grounded air inlet ducts, or use groundwater to circulate in building elements to reduce heat radiation.

A two-courtyard building: air from the shaded courtyard flows over evaporative coolers to the larger and warmer courtyard. A courtyard pool has a beneficial effect during the day and night.

Evaporative cooling

Air passing over water will cause evaporation, and as a result of this process heat is absorbed and the air is cooled. The evaporated water is retained in the air, thus increasing its humidity; for this reason evaporative cooling is suitable for dry climates.

The simplest principle can be achieved by channelling breezes over pools or water sprays before they enter a building. To ensure that the cooled and humidified air enters the building, the pool should be placed between walls such as in a courtyard. A spray pond is more effective than a still pond, and has the additional advantage that it will not only cool the air but also clean it, as the dust particles in the air will stick to the water droplets.

As discussed previously, evaporative cooling can be incorporated into wind towers or as part of openings.

Green plants will also have an effect due to the evaporation of water, but also act as purifiers of the air and as suppliers of oxygen.

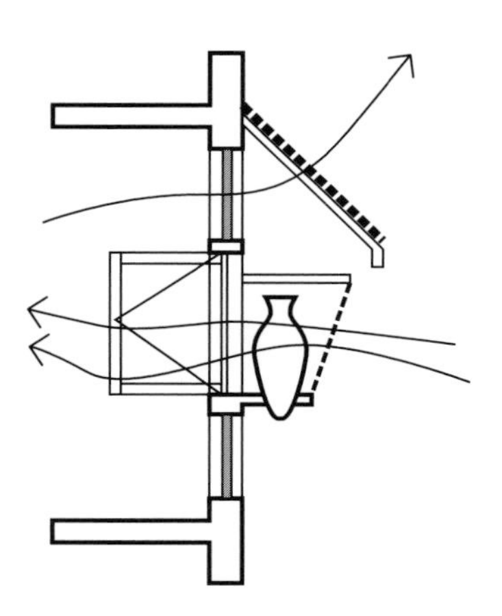

Air is cooled as it passes over the surface of a water-filled earthenware pot. Based on a drawing by Development Workshop.

Forced evaporative coolers

The potential of direct evaporative coolers will depend on the potential of the air to absorb humidity. The drier the air, the greater is the cooling potential as a greater amount of water can be evaporated and heat absorbed. This method is therefore well suited to hot dry climates.

A method of forced evaporative cooling in a simple form for use in small rooms would be evaporation from a water-filled earthenware pot under an electric fan.

Forced evaporative coolers provide a reliable and flexible method of cooling using a fan to create a controllable and constant source of air movement. Water is pumped from a small reservoir at the base to an upper tank, from which it is allowed to trickle slowly over matting placed across the opening. This cooler can be combined with solar cells to power the system. In such a cooling system with an external temperature of 35°C and a relative humidity of 40%, the air can be cooled by 5°C; at a relative humidity of 10% a cooling of the air by 11°C is possible.

The amount of water consumption needed for evaporative coolers has been queried because the method of cooling is used in dry zones where water resources are normally limited. For an ordinary 'desert cooler', as described above, the water consumption for cooling 100 m^3 of air by 5°C is up to 15 litres/h (from *Climate Responsive Building*, by Paul Gut and Dieter Ackerknecht). Less water-consuming evaporative cooling systems

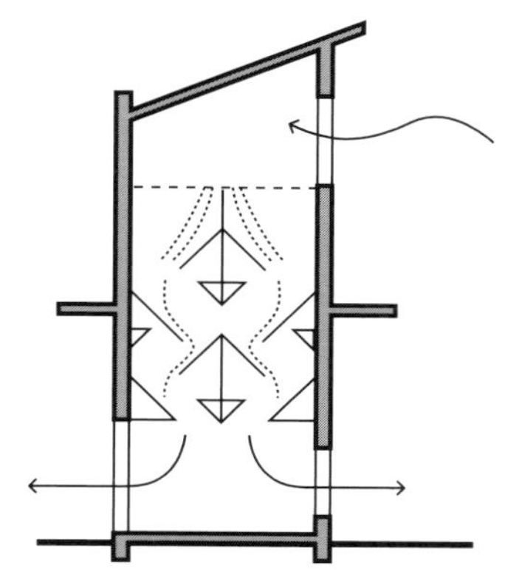

An evaporative cooler combined with a wind tower. A mat of wetted charcoal can be placed in wind towers to increase the surface area across which evaporation can take place. Piped water can be used to keep the mats moist. A pipe with a series of small holes can be placed above an opening or, as in this case, at the top of the wind tower. Based on a drawing by Development Workshop.

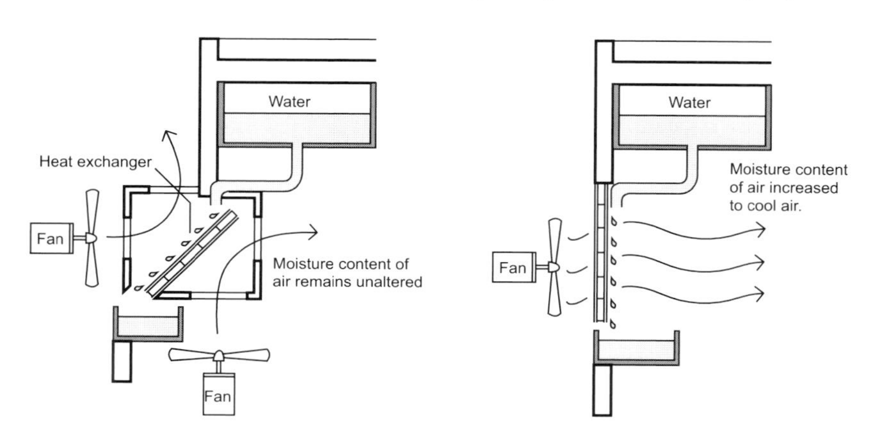

(left) Example of an indirect evaporative cooling system.
(right) Example of a direct evaporative cooling system.

The Botswana Technology Centre in Gaborone; see also illustrations on page 67. The building has a highly efficient evaporative cooling system, as mentioned in the text. The cooled air is mechanically and continuously distributed to the hollow core concrete slabs to cool the structure and to the occupied spaces to cool the ambient air (3–6 air changes per hours). (left) In low pressure areas/niches in the internal street the hot air from the office spaces is exhausted in a tube system. The air suction system is driven by the stack effect and by wind cowls on the roof (right).

have been developed using spray humidification of the air. The BOTEC building in Botswana, seen in several of the illustrations in this Guide, uses a mixture of compressed air and water to create a humidifying mist, which is blown into the incoming air stream via nozzles. According to surveys of the system, it uses 85.5 litres/h to cool an air volume of 6 m^3/s by 5°C. It is about 35 times more 'water efficient' than the desert cooler. The water used for evaporative cooling of the building is supplied from rainwater collection tanks.

In warm humid regions the potential use of a direct evaporative cooler is small owing to high humidity levels. In this case an indirect evaporative cooler that relies on a heat exchanger can be used in most cases. This will allow air to be cooled without increasing humidity levels in the internal space.

Other evaporative cooling methods

Additional evaporative cooling methods for the hot dry zone are the spraying of water onto roofs and walls or the introduction of a pond on the roof which acts as a thermal storage/buffer 'tank'. These cooling methods will reduce the temperature of the roofs and walls and delay or eliminate radiation towards the interior. These systems have been studied by B. Givoni, and are demonstrated in his book *Passive and Low Energy Cooling of Buildings*. They have had only limited application.

Earth cooling

Soil temperature below the surface is found to be equal to average annual air temperature, and is therefore considerably lower during the hot season than air temperatures. To minimise the impact of heat on the topsoil adjacent to the building it should be surrounded with plants, which will absorb most of the solar radiation so that only a small amount will be absorbed by the earth.

The building envelope in full or partial contact with the soil provides cooling owing to conduction between the soil and the wall, which subsequently radiates cold to the internal space and at the same time absorbs the internal heat by convection. Earth cooling (or heating) represents the ultimate application of a thermal mass to the building. The potential for large-scale construction of such buildings is limited. This is due mainly to rather high construction costs and difficult daylighting conditions. However, construction of a cellar through which the incoming air passes before entering the upper floors is an earth-cooling method that has been known for centuries.

An earth-cooling method that is currently being practised draws air through tubes or passages below the ground to cool the air before entering a building. Humidification of the air at the intake for the ground tubes can further reduce the temperature of the air. Attention should be paid to possible condensation in the tubes in warm humid regions/periods.

A thermal storage reservoir constructed underground and made of materials with thermal storage capacity through which the incoming hot air passes and is cooled on its way to the interior of the building is another type of earth cooling. In hot dry regions where winter night temperatures drop below zero, the system also offers heating possibilities when the air circulates from the thermal reservoir to the interior of the building.

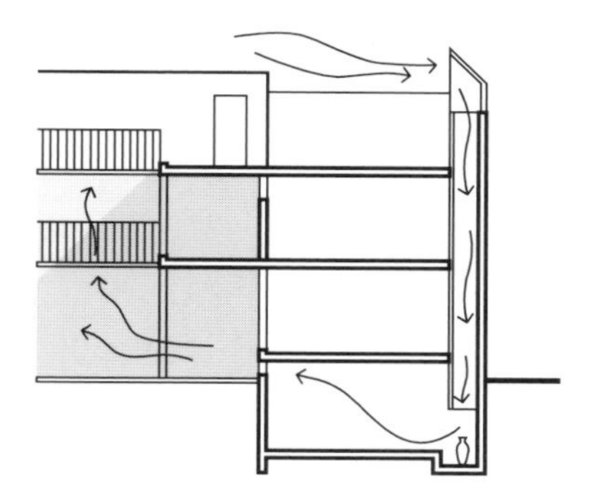

Cooling of the incoming air takes place when it passes through an earth-cooled cellar.

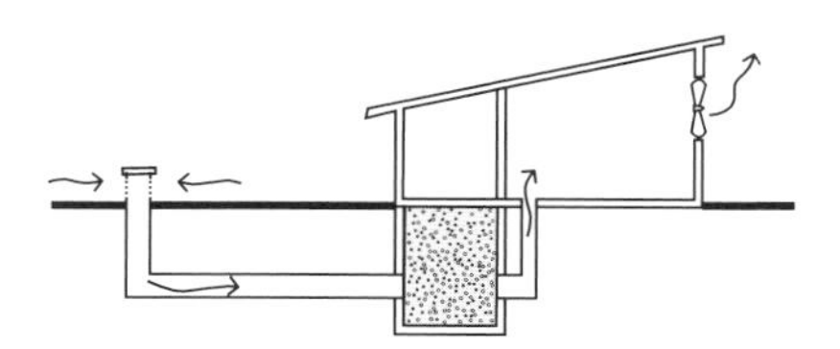

Earth cooling from an underground thermal storage reservoir, e.g. a rock chamber. Essentially, the system cools the incoming air during day, it having passed through the underground tubes and the rock chamber; during night the air cools the rocks. The system has been used in several places; the illustration shows the system on the Harare International School, Zimbabwe. Architects: Pearce Partnership. Engineers: Ove Arup & Partners.

The National Food Technology Research Centre in Kanye, Botswana. The building uses solar chimneys to exhaust hot air and ground pipes with moistened air to cool incoming air. Architects: Erik S. Leus, Brian Marland and Bena Tibe. Engineers: Lasco Engineering. (above right) Part of the building facade with the solar chimneys and a very sophisticated shading system. (above) The air intake structure with a water fountain.

Notes: Radiative cooling methods

A cooling and heating method that was developed in Europe in the 1950s uses circulated piped water in ceilings, floors and walls. A similar system with, for example, circulated piped groundwater, which is naturally colder than the environment in the hot season, is a possible option for reducing heat radiation from building elements in hot dry climates. The system may also be used in warm humid regions with very high relative humidity, but attention must be paid to possible condensation by controlling the surface temperature. The reduced surface temperature can, together with possible air movements, significantly reduce the temperature experienced by the occupants.

Air conditioning and combinations of systems

The term 'air-conditioning systems' refers to the conventional ducted systems used to control the input and output of air as well as the quality of air in a building. The use of this system may be unavoidable, for example when the requirements are for constant cooling at a very high rate, or where there is a need for a quick response to variations in ventilation and cooling demands of spaces in buildings.

Decisions on how a building's comfort is to be maintained need to be made at the beginning of the design process. It is not necessarily a question of either active or passive measures; a combination is also an option where natural means are used for certain parts of the year/day and active means are used during other parts of the year/day. A building can also be zoned for different internal requirements, and a mixed system can be introduced for simultaneous use of natural ventilation and natural cooling as well as air conditioning. In each case, appropriate building designs and management systems have to be developed.

The use of air conditioning will require a different building construction from that needed if the climate itself is the driving force used to satisfy comfort levels. Certain design aspects, for example thermal storage capacity, thermal insulation and its locations in the construction, call for a different approach to minimising the energy required to maintain an artificially cool internal environment.

The design approach is in fact parallel to that used in maintaining a warm internal climate in cold regions. For combined or mixed systems, the design issues becomes more complicated owing to mixed conditions and measures to counteract the external climate.

Conscious management and proper maintenance of air-conditioning units or systems are important in order to reduce the consumption of energy.

It should be noted that to introduce energy-consuming systems to cool and ventilate a building also requires a constant power supply, which cannot always be secured in many developing countries. If an unreliable power supply is in place, the added expense of a back-up system will also be required.

Design principles: natural ventilation and cooling

Hot dry zones

- External surrounding features and a building's siting in relation to these should maximise the use of available breezes and ensure that the air is cooled and dust free before it enters a building.

- Ventilation of internal spaces needs to be controlled. Principally, during the hot periods of the day ventilation should be minimised. Night-time structural cooling is required to remove heat stored in a building's structure. This can be satisfied by opening up the structure at night, and requires that air movements be directed to pass over warm surfaces (e.g. the underside of ceilings) and/or it can be designed to cool a specific area such as roof or wall cavities. Air movements can also be directed towards occupants of internal spaces for cooling when the temperature of the air is well below the body temperature.
- Ventilation openings should be designed to be flexible in operation, depending on changing external conditions. They can be located at differing heights to use the stack effect and be of differing sizes to maximise the pressure/suction effect. The same openings can be regulated, by opening and closing, to utilise different effects.
- A variety of stimulated natural ventilation structures and devices, for example wind towers and/or roof devices, can be used to enhance natural ventilation. They can also be designed to operate on different effects and be assisted by simple mechanical ventilation, depending on changing external conditions, e.g. dust, wind direction.
- Evaporative cooling can be used to cool both a building's structure and its internal spaces; this involves humidifying the air before it enters the building. Evaporative cooling methods can be used in stimulated natural ventilation structures and devices as part of an opening or incorporated in the overall landscape strategy, e.g. fountains, water pools of a building. Direct evaporative coolers can be used to cool the incoming air.
- Ventilation of wall structures and earth cooling methods can be used to reduce heat gains from the building envelope as well as for cooling the incoming air.

Warm humid zones

- External features and a building's siting in relation to these should be designed to encourage maximum air movement.

Internal features such as walls and partitions should be designed to minimise impediments to air movement.

- Ventilation of internal spaces needs to be continuous throughout the day and night to avoid the storage of heat by the structure. This will require that the structure be designed to be open to capture all available breezes.
- Openings should be large, to maximise cross-ventilation of internal spaces. During periods when wind speeds are low, openings can be designed to utilise the stack effect. They can also be designed to ventilate specific structural elements, for example a roof cavity.
- In regions where the temperature at midday in the hot season reaches an uncomfortable level, and stimulated natural ventilation cannot create a comfortable internal environment, the building envelope should be closed to avoid the entrance of the external air, and passive systems for cooling of the air and/or the structure could be introduced.
- Simple mechanical devices can be used to supplement natural ventilation. When external wind speeds cannot satisfy the ventilation and cooling requirements, they can be used to generate internal air movement, improve air distribution and increase air velocities. Fans can be located and designed to operate in a specific area or at a specific time. They can be directed and used to increase air movement at varying heights and areas to increase comfort conditions.
- Evaporative cooling can be used to cool internal spaces, but more importantly to cool the body; this involves directing air movement at occupancy level. This requires that openings, shading devices, etc. should be designed and located accordingly, and they should be flexible enough to respond to changing external conditions. Indirect evaporative coolers can be used to cool the incoming air.
- Ventilation of wall structures and earth cooling methods can be used to reduce heat gains from the building envelope as well as to cool the incoming air.

Appendix A:

Active design process checklist

Design

Siting

- Consider the optimum microclimate effects from site design
- Minimise exposure to solar radiation
- Consider exposure to cooling breezes

The built environment

- Consider the grouping of buildings
- Optimise orientation for solar protection and wind direction
- Consider the density and surface to volume ratio of buildings
- Consider building depth and spacing
- Consider urban landscaping and vegetation as well as pedestrian networks

The external environment

- Ensure that external spaces are comfortable (e.g. shaded) and landscaped to reduce heat gains and maximise air flow through buildings
- Design landscaping that assists in microclimate management to reduce energy use

Internal planning

- Plan and zone according to function, occupancy levels and time of use
- Plan to minimise the number of different thermal zones
- Identify particular functions or uses requiring special conditions, control or monitoring; deal with them separately or in zones rather than raising the servicing and energy consumption of the building as a whole
- Plan for flexibility for future changes

Thermal response of building envelope

- Design openings to reduce solar gains, maximise daylighting and air flow
- Select materials, construction and finishes to give optimum thermal response and reflect occupancy patterns
- Design the building envelope and components to make thermal control as easy as possible
- Maximise heat removal through the envelope by natural ventilation and cooling aids
- Provide ventilation provisions that are climate responsive and reflect the time of use, function and occupancy levels of a building
- Consider appropriate air movement patterns
- Openings designed for ventilation purposes: consider positioning, size, treatment and use

Controls

- Building envelope that allows control and adaptability in response to climatic variations
- Openings: operable to control internal air flow
- Protection of openings: flexible to regulate heat gains and daylighting levels depending on the position of the sun

Allowance for future changes

- Internal planning
- Building envelope
- Service installation

Overall checks

- Designers should check all decisions made throughout the design process and go back and recheck earlier decisions
- Use checking tools to confirm design decisions

Details

External openings

- Location, sizing and treatment that will maximise ventilation and minimise solar heat gains
- Location, sizing and treatment to maximise natural lighting and minimise glare
- The provision of shading devices and other forms of sun protection
- The provision of air inlet and outlet openings in the roof cavity and walls for heat dissipation

Building envelope

- Materials and construction selection: climate responsive and control heat entry and removal
- Consider all potential areas where air gaps and thermal bridging may occur

Ventilation and cooling provisions

- Check zoning
- Check that the building envelope is designed to optimise ventilation and cooling possibilities
- Select and locate appropriate ventilation and cooling devices and systems so that they respond to specific and changing climate conditions
- Provisions for appropriate operation: user control or management control

- Mechanical ventilation should be used to compensate only for climate impacts rather than poor building design

Control internal heat sources
- Minimise casual internal heat gains from lighting, appliances and occupancy to avoid internal overheating
- Artificial lighting should be zoned, controlled and activated in areas where and when there is a need for light
- Minimise heat release from appliances (e.g. refrigerators) and other equipment (e.g. computers) by selection made on the basis of energy efficiency
- Minimise heat gains from occupants; consideration at the design stage of possible occupancy levels of each room and the range of activities
- Adequate ventilation of densely occupied spaces
- Consider zoning/grouping high-grade heat sources for better energy management

Lighting installation
- Use minimum illumination levels and consider task lighting rather than general illumination
- Use energy-efficient lamps and luminaires
- Consider time and intensity controls
- Use light-coloured internal finishes to improve lighting conditions

Design detailing and documentation

- Ensure that appropriate materials are specified and construction is properly dimensioned to avoid oversizing
- Ensure that documentation states all requirements for commissioning, testing, handover, occupation and maintenance
- Provide user education through manuals/training
- Allow for the potential of future changes to the building

Design feedback

Designers should return to learn the opinions of occupants, observe performance and to take spot measurements where appropriate: to adjust building performance and for future design reference

Design and checking tools

There are various tools at the disposal of a designer that can be used to check design decisions. Their use may range from simple methods for broad and initial design decisions to the use of complex computer modelling programs to check specific design solutions. Regardless of which tools are used, the designer should be aware that they are not capable of solving problems in all design situations and should rather be viewed as design aids. A few examples of these methods are outlined below.

Simple analysis

Solar charts provide a comprehensive and easy to use design aid for the prediction of solar radiation exposure and shading of a building. Solar charts represent the sun–building relationship in a graphic form. They can be used as a checking tool for a proposed solution, but are primarily intended as a design tool.

Mahoney tables are used to arrive at a series of performance specifications. The input information required for the tables is the various climate data for the intended building location. This can be used to develop a series of climate indicators that can be translated into performance specifications or sketch design recommendations for simple buildings. This method is also available as a computer program. In this case the only data necessary are temperature, humidity, rainfall and geographical location. The computer program establishes a preferred comfort range, carries out a diagnosis, and translates this into simple elementary recommendations.

Physical modelling

Wind tunnel test models are physical models of buildings on a site that can be used in conjunction with wind tunnels to examine air flow around buildings. Test results may lead to modifications of basic features such as the volume, shape or the siting of a building. The need for such testing arises in the case of larger developments or schemes involving groups of buildings. In this situation it may be useful to test the effect of upwind objects on air flow reaching the proposed building or the effect of the proposed building on the existing environment in the downwind direction. Wind tunnels can also be used to test air flow through a particular space. This will require a large-scale model that has a true representation of openings and objects within the space. In both cases the testing can take three forms:
- visualisation of air flow patterns, using smoke traces;
- measurement of pressure differences;
- measurement of air velocity at various points.

Saline test models show temperature gradients and air flow. For instance, they can be used to determine heat gains from internal heat sources such as people, computers and appliances. A physical model with adjustable openings made of transparent plastic is required. The model is totally immersed upside down in a large tank of cold fresh water, and the heat gains are modelled by injecting a saline solution of specific density at the heat source. Since salt water is heavier than fresh water, the saline solution falls through the model, and the rate of mixing with the fresh water causes a change of density that is numerically similar to that of temperature diffusion in air. The temperature at any point is given by a measurement of density, and the flow rate is measured by physical flow. Coloured dyes can be added to the saline solution so that the process can be easily seen. This can also be used to test the air flow through openings due to temperature differences. An illustration of a saline model is shown on page 19.

Computer modelling

A number of computer programs are available to predict design performance. The capacity of the programs is becoming more and more advanced, which will make future physical modelling (mentioned above) less necessary. They range from rather simple calculations requiring general data input to sophisticated modelling techniques requiring large amounts of exact data. Below is a brief list of the possible outputs that can be obtained from computer modelling:

- computerised solar charts for various latitudes and times of the year and day;
- shading calculation of buildings and elements incorporated into the building envelope;
- design of openings: calculations for daylighting and air movement through openings;
- the thermal performance of materials and elements;
- the thermal performance of the building envelope as a whole or of particular facades;
- internal heat flow calculations through the building or a particular space;
- internal air flow predications around or through a building or in a particular space;
- performance assessment of natural ventilation or cooling devices;
- humidity flow through building elements.

Appendix B:

Rehabilitation guide for existing urban, external and building environments

When making rehabilitation decisions, certain conditions cannot be altered, for example the siting and orientation of buildings; others can prove costly or difficult to change, e.g. the size and positioning of openings. However, individual components in the urban and building environment can be altered and modified to reflect specific climatic characteristics.

Rehabilitation initiatives can be adopted to alter multiple conditions. For instance, the implementation of landscaping strategies can be used to rehabilitate and improve both external and, indirectly, internal conditions at the urban and building scale. Rehabilitation can also be target specific, such as fitting shading devices over existing openings.

Each decision regarding rehabilitation should be done in relation to specific climatic conditions, the changed function or the time of use or occupancy levels of internal and external spaces and in relation to how these decisions will impact on parts that remain unaltered.

Rehabilitation possibilities that can be incorporated at various design levels are presented below.

Note:
Abbreviations HD (hot dry) and WH (warm humid) refer to recommendations specific to that climatic zone. The remaining recommendations are general.

Considerations

Siting

- The siting of a building is generally a given factor. However, the siting of other elements can be used to affect the microclimatic conditions of an existing building.
- The siting of shelter belts and manipulation of land contours can be used to alter wind forces in both direction and velocity.
- If the rehabilitation of an existing building involves the addition of new buildings, they can also be sited to alter the microclimate that surrounds existing buildings, e.g. their exposure to solar radiation and air movement.

The built environment

Orientation

- Orientation of an existing building cannot be altered, but measures can be made to design according to orientation when a building is being rehabilitated.
- For change of use of a building: the function and use of internal spaces should be related to orientation. For example, rooms used in the afternoon and evening can be located facing in an easterly direction.
- For addition of components to the building envelope: the design of additions and alterations made to the building envelope should be related to orientation.

Urban environment

- Pedestrian networks: the addition of pedestrian covered walkways, planting provisions and provision of cool shady rest spots.
- Existing urban spaces can be designed to act as microclimatic controls through landscaping strategies, planting and water provisions.
- Building facades can be designed to provide urban-scale improvements such as covered colonnades and awnings.

Built environment

- Examine whether the surface to volume ratio and the density are suited to the climatic condition. If new buildings are part of a rehabilitation programme, they can be used to achieve compactness (HD) or openness (WH).
- Secondary elements: projections can be added to a building to create shaded transitional spaces such as verandas and colonnades to reduce heat gains made by the building.
- Incorporate landscape strategies that can alter the surrounding condition of a building: planting to shade a building, a canopy selection that does not restrict air movement, inclusion of water for evaporative cooling (HD) and ground cover selection to reduce heat reflection and glare, for example grass instead of paving.

The external environment

- Rehabilitation of existing external spaces to assist in microclimate management at urban and building scale.

- Water, planting and enclosure devices such as walls can be used to create pleasant and usable outdoor spaces (HD).
- Enclosing spaces so that they can be managed and, by alteration of microclimate, improve air flow (HD).
- Extending the building envelope such as the roof overhangs and by using verandas to act as transitional and protected outdoor spaces (WH).
- Planting: tall trees with high canopies, to provide shade but which do not restrict air flow around and through buildings (WH).
- Rain protection devices can be incorporated: planting or structures (WH).

Building envelope

Roof

- Roof forms can be rehabilitated to reflect internal functions, orientation and climate characteristics.
- Investigate the potential use of composite or two separate roof structures such as a flat roof converted to a composite roof with a cavity, or a light roof structure added above an existing roof, to reduce heat transfer (HD).
- Investigate the possibility of the addition of a ceiling to create a composite structure.
- Implement a landscaping strategy that directs air flow across the external surface of the roof for cooling.
- Investigate the possibility of installing wind-catching devices to cool the roof cavity or wind-catching structures to cool internal spaces, and use wind-catching devices or structures that can be adjusted to suit changing climate conditions.
- Investigate the possibility of including air inlet and outlet openings to cool a roof cavity or the underside of the roof, sized and positioned to maximise the potential of prevailing winds.
- Select materials based on climate characteristics, function and time of use of the internal spaces and according to how they will affect the existing conditions.

Walls

- The rehabilitation of walls should reflect the function and use of rooms adjacent to them and their orientation.
- Investigate the possibility of rehabilitating an existing wall into a composite wall as a means of reducing solar radiation impact (HD).
- Investigate the potential of incorporating ventilation openings in new walls to act as a cooling mechanism.
- Investigate the incorporation of secondary wall elements to reduce solar radiation impact such as verandas and colonnades.
- Reduce heat gains (and glare) affecting the walls by a landscaping strategy such as shade planting and groundcover selection.
- Select materials based on climate characteristics, orientation, function and time of use of the internal adjacent spaces and study how they will affect the existing conditions.

Floors

- Selected materials to improve internal comfort conditions such as hard cool surfaces (HD); light materials and openings (WH).

Openings

- It may prove difficult or costly to change the size and positioning of openings in existing buildings. If such decisions are made the openings should be designed according to the recommendations for opening design made in this Guide.
- The addition of shading and protection devices for existing openings offers a real potential to reduce heat gains and to increase or to direct internal air movement, to reduce glare and maximise internal daylighting.
- Shading devices placed on the outside of openings are the first line of defence against heat gains through openings.
- Shading devices such as louvres can be adjusted and can alter the direction of air flow and lighting.
- Select materials for shading devices that will reflect heat rather than be an added heat source.
- Select glass that can reduce heat gains by absorbing or reflecting heat.

Air gaps and thermal bridges

- Investigate the whole of the building to ensure that there is no uncontrolled heat entry (HD) from air gaps and infiltration.
- Ensure that all new building components introduced (e.g. the alteration of a single wall into a composite wall) do not create thermal bridges.

Materials

- All new materials or elements that are to be part of a rehabilitation programme should be analysed in terms of their thermal properties. Choices should reflect climate characteristics, orientation, internal building functions and how they react in relation to existing materials and building elements.

Internal planning

- Plan and zone according to new building: function, occupancy levels and time of use.
- Plan and arrange internal spaces in relation to external facades and their orientation.
- Plan to minimise the number of different thermal zones.
- Plan and zone spaces with particular functions or requirements to reduce energy use or new service requirements.
- Ensure that partitions do not impede air movement such as for ventilation.
- Ensure that rooms used during the day are related to external facades to maximise daylighting.

Internal building provisions

- Refer to Appendix A: Control internal heat sources and lighting installation.

Ventilation and cooling provisions

- Ensure that appropriate ventilation levels are provided to reflect climate characteristics, functions and time of use and occupancy levels of individual internal spaces.
- Incorporate structures or devices for structural cooling and ventilation purposes (HD).
- Regulate openings to encourage the stack effect.
- Ensure that internal partitions do not impede internal air flow.
- If new buildings are to form part of a rehabilitation strategy, ensure that they are sited to encourage air movement to create desired air flow patterns.
- If new building elements are introduced, investigate the possibility of combination solutions such as use of a staircase to create stack and venturi effects.
- Investigate the need for a means of supplementary natural ventilation or simple mechanical devices that can be used when wind speeds are low and for structurally cooling components of the building envelope.
- Investigate the use of solar-powered systems to run mechanical devices.
- Incorporate ventilation devices that direct air movement across desired surfaces such as hot internal surfaces or across the body.
- Use landscaping strategy for evaporative cooling of external and internal spaces (HD).

Appendix C:

Solar charts

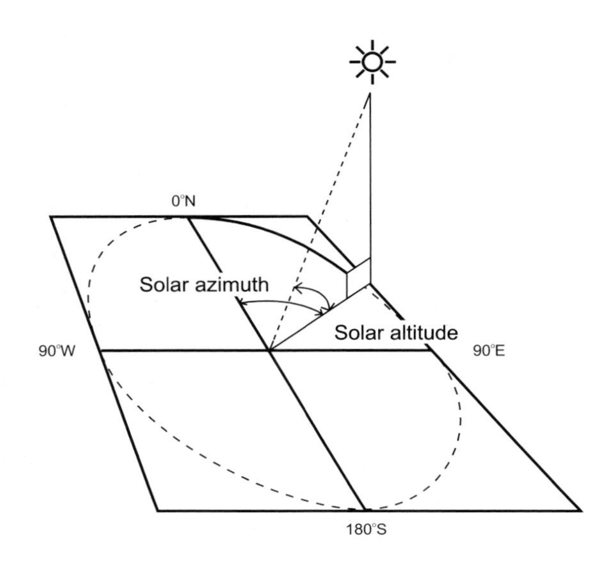

A visual explanation of solar angles

The sun's position in the sky can be identified by two angles.

Solar altitude angle: This describes how high the sun appears in the sky. The angle is measured between an imaginary line drawn between the observer and the sun and the horizontal plane on which the observer is standing.

Solar azimuth angle: This is the angular distance between due north and the projection of the line of sight to the sun on the ground.

These two angles can be found for any day of the year and any hour of the day from solar charts or computer programs, and give the geometry of a shadow from an overhang or solar protection construction.

This Appendix provides solar angles for 21 March, 21 June, 21 September and 21 December between the hours 6:00 and 18:00 for latitudes 30° north and south of the equator at 5° intervals. The sun is in the northern hemisphere from 21 March to 21 September, and the extreme position north is found on 21 June. 21 December is the date of the extreme position south of the sun, which is found in the southern hemisphere from 21 September to 21 March.

To maintain the validity of the solar charts, the hour lines are not indicated with the time as shown on a clock. Local time does not, as a rule, coincide with true *local solar time*. The proper designation of the hour line is determined by the difference between the time in a country (i.e. standard time) and local solar time. True local solar midday is the moment when the sun reaches its highest altitude, and the azimuth is at exactly 180° or 0° (according to location and season).

The following solar charts have been generated using the Sunpath* computer program with minor corrections based on information from the Danish Technical University.

** Produced by Dr Ross McCluney, Florida Solar Energy Center*

titude 0° North

	21 March / 21 September		21 June		21 December	
Solar time	Solar altitude	Solar azimuth	Solar altitude	Solar azimuth	Solar altitude	Solar azimuth
06:00			1	67	1	113
07:00	15	90	14	66	14	114
08:00	30	90	27	64	27	117
09:00	45	90	41	59	41	122
10:00	60	90	53	49	53	131
11:00	75	90	62	31	62	149
12:00	90	0	65	0	67	180
13:00	75	90	62	31	62	149
14:00	60	90	53	49	53	131
15:00	45	90	41	59	41	122
16:00	30	90	27	64	27	117
17:00	15	90	14	66	14	114
18:00			1	67	1	113

titude 5° North

	21 March / 21 September		21 June		21 December	
Solar time	Solar altitude	Solar azimuth	Solar altitude	Solar azimuth	Solar altitude	Solar azimuth
06:00	0	90	2	67		
07:00	15	91	16	67	12	115
08:00	30	92	30	66	25	119
09:00	45	94	43	62	38	125
10:00	60	97	56	55	49	135
11:00	74	106	67	37	58	153
12:00	86	180	72	0	62	180
13:00	74	106	67	37	58	153
14:00	60	97	56	55	49	135
15:00	45	94	43	62	38	125
16:00	30	92	30	66	25	119
17:00	15	91	16	67	12	115
18:00	0	90	2	67		

titude 10° North

	21 March / 21 September		21 June		21 December	
Solar time	Solar altitude	Solar azimuth	Solar altitude	Solar azimuth	Solar altitude	Solar azimuth
06:00	1	90	4	67		
07:00	15	92	18	68	10	116
08:00	30	95	32	69	23	121
09:00	44	99	45	67	35	128
10:00	59	106	58	61	46	139
11:00	72	121	70	45	54	156
12:00	81	180	77	0	57	180
13:00	72	121	70	45	54	156
14:00	59	106	58	61	46	139
15:00	44	99	45	67	35	128
16:00	30	95	32	69	23	121
17:00	15	92	18	68	10	116
18:00	1	90	4	67		

Latitude 15° North

	21 March / 21 September		21 June		21 Decemb	
Solar time	Solar altitude	Solar azimuth	Solar altitude	Solar azimuth	Solar altitude	Solar azimu
06:00	1	90	6	67		
07:00	15	94	20	70	7	117
08:00	30	98	33	72	20	122
09:00	43	104	47	73	32	130
10:00	57	113	61	71	42	142
11:00	69	133	74	63	49	159
12:00	76	180	82	0	52	180
13:00	69	133	74	63	49	159
14:00	57	113	61	71	42	142
15:00	43	104	47	73	32	130
16:00	30	98	33	72	20	122
17:00	15	94	20	70	7	117
18:00	1	90	6	67		

Latitude 20° North

	21 March / 21 September		21 June		21 Decemb	
Solar time	Solar altitude	Solar azimuth	Solar altitude	Solar azimuth	Solar altitude	Solar azimu
06:00	1	90	8	68		
07:00	14	95	21	72	5	117
08:00	28	101	35	75	17	124
09:00	42	108	48	77	28	133
10:00	55	120	62	77	38	145
11:00	66	141	76	77	44	161
12:00	71	180	87	0	47	180
13:00	66	141	76	77	44	161
14:00	55	120	62	77	38	145
15:00	42	108	48	77	28	133
16:00	28	101	35	75	17	124
17:00	14	95	21	72	5	117
18:00	1	90	8	68		

Latitude 25° North

	21 March / 21 September		21 June		21 December	
Solar time	Solar altitude	Solar azimuth	Solar altitude	Solar azimuth	Solar altitude	Solar azimu
06:00	1	90	10	69		
07:00	14	96	23	74	3	118
08:00	27	103	36	78	14	125
09:00	40	112	49	82	25	134
10:00	52	125	63	87	36	147
11:00	62	147	76	93	40	162
12:00	66	180	88	180	42	180
13:00	62	147	76	93	40	162
14:00	52	125	63	87	36	147
15:00	40	112	49	82	25	134
16:00	27	103	36	78	14	125
17:00	14	96	23	74	3	118
18:00	1	90	10	69		

atitude 30° North

	21 March / 21 September		21 June		21 December	
Solar time	Solar altitude	Solar azimuth	Solar altitude	Solar azimuth	Solar altitude	Solar azimuth
06:00	1	90	12	69		
07:00	13	97	24	76	1	118
08:00	26	106	37	82	12	126
09:00	38	116	50	88	21	136
10:00	49	130	63	97	29	148
11:00	57	151	75	113	35	163
12:00	61	180	83	180	37	180
13:00	57	151	75	113	35	163
14:00	49	130	63	97	29	148
15:00	38	116	50	88	21	136
16:00	26	106	37	82	12	126
17:00	13	97	24	76	1	118
18:00	1	90	12	69		

atitude 0° South

	21 March / 21 September		21 June		21 December	
Solar time	Solar altitude	Solar azimuth	Solar altitude	Solar azimuth	Solar altitude	Solar azimuth
06:00			1	67	1	113
07:00	15	90	14	66	14	114
08:00	30	90	27	64	27	117
09:00	45	90	41	59	41	122
10:00	60	90	53	49	53	131
11:00	75	90	62	31	62	149
12:00	90	0	65	0	67	180
13:00	75	90	62	31	62	149
14:00	60	90	53	49	53	131
15:00	45	90	41	59	41	122
16:00	30	90	27	64	27	117
17:00	15	90	14	66	14	114
18:00			1	67	1	113

Latitude 5° South

	21 March / 21 September		21 June		21 December	
Solar time	Solar altitude	Solar azimuth	Solar altitude	Solar azimuth	Solar altitude	Solar azim
06:00	0	90			2	113
07:00	15	89	12	65	16	113
08:00	30	88	25	61	30	114
09:00	45	86	38	55	43	118
10:00	60	83	49	45	56	125
11:00	74	74	58	27	67	143
12:00	86	0	62	0	72	180
13:00	74	74	58	27	67	143
14:00	60	83	49	45	56	125
15:00	45	86	38	55	43	118
16:00	30	88	25	61	30	114
17:00	15	89	12	65	16	113
18:00	0	90			2	113

Latitude 10° South

	21 March / 21 September		21 June		21 December	
Solar time	Solar altitude	Solar azimuth	Solar altitude	Solar azimuth	Solar altitude	Solar azimu
06:00	1	90			4	113
07:00	15	88	10	64	18	112
08:00	30	85	23	59	32	111
09:00	44	81	35	52	45	113
10:00	59	74	46	41	58	119
11:00	72	59	54	24	70	135
12:00	81	0	57	0	77	180
13:00	72	59	54	24	70	135
14:00	59	74	46	41	58	119
15:00	44	81	35	52	45	113
16:00	30	85	23	59	32	111
17:00	15	88	10	64	18	112
18:00	1	90			4	113

Latitude 15° South

	21 March / 21 September		21 June		21 December	
Solar time	Solar altitude	Solar azimuth	Solar altitude	Solar azimuth	Solar altitude	Solar azimu
06:00	1	90			6	113
07:00	15	86	7	63	20	110
08:00	30	82	20	58	33	108
09:00	43	76	32	50	47	107
10:00	57	67	42	38	61	109
11:00	69	47	49	21	74	117
12:00	76	0	52	0	82	180
13:00	69	47	49	21	74	117
14:00	57	67	42	38	61	109
15:00	43	76	32	50	47	107
16:00	30	82	20	58	33	108
17:00	15	86	7	63	20	110
18:00	1	90			6	113

atitude 20° South

	21 March / 21 September		21 June		21 December	
Solar time	Solar altitude	Solar azimuth	Solar altitude	Solar azimuth	Solar altitude	Solar azimuth
06:00	1	90			8	112
07:00	14	85	5	63	21	108
08:00	28	79	17	56	35	105
09:00	42	72	28	47	48	103
10:00	55	60	38	35	62	103
11:00	66	39	44	19	76	103
12:00	71	0	47	0	87	180
13:00	66	39	44	19	76	103
14:00	55	60	38	35	62	103
15:00	42	72	28	47	48	103
16:00	28	79	17	56	35	105
17:00	14	85	5	63	21	108
18:00	1	90			8	112

atitude 25° South

	21 March / 21 September		21 June		21 December	
Solar time	Solar altitude	Solar azimuth	Solar altitude	Solar Azimuth	Solar altitude	Solar Azimuth
06:00	1	90			10	111
07:00	14	84	3	62	23	106
08:00	27	77	14	55	36	102
09:00	40	68	25	46	49	98
10:00	52	55	36	33	63	93
11:00	62	33	40	18	76	87
12:00	66	0	42	0	88	0
13:00	62	33	40	18	76	87
14:00	52	55	36	33	63	93
15:00	40	68	25	46	49	98
16:00	27	77	14	55	36	102
17:00	14	84	3	62	23	106
18:00	1	90			10	111

atitude 30° South

	21 March / 21 September		21 June		21 December	
Solar time	Solar altitude	Solar azimuth	Solar altitude	Solar azimuth	Solar altitude	Solar azimuth
06:00	1	90			12	111
07:00	13	83	1	62	24	104
08:00	26	74	12	54	37	98
09:00	38	64	21	44	50	92
10:00	49	50	29	32	63	83
11:00	57	29	35	17	75	67
12:00	61	0	37	0	83	0
13:00	57	29	35	17	75	67
14:00	49	50	29	32	63	83
15:00	38	64	21	44	50	92
16:00	26	74	12	54	37	98
17:00	13	83	1	62	24	104
18:00	1	90			12	111

Bibliography

Main references for this book, recommended for further studies

[In square brackets, the chapter of the present work in which these have been principally used.]

Allard, Francis (ed.), *Natural Ventilation in Buildings: A Design Handbook*, James & James Ltd, London, 1998. [Chapter 7]

Asimakopoulos, D. and Santamouris, M. (eds), *Passive Cooling of Buildings*, James & James Ltd, London, 1996. [Chapter 7]

Battle McCarthy, Consulting Engineers, *Wind Towers*, Academy Editions (UK), 1999. [Chapter 7]

Clements-Croome, Derek (ed.), *Naturally Ventilated Buildings: Buildings for the Senses, the Economy and Society,* E & FN Spon, London, 1997. [Chapter 3]

Evans, Martin, *Housing, Climate and Comfort*, Architectural Press Ltd, London, 1980. [Chapters 1–7]

Givoni, Baruch, *Man, Climate and Nature,* Applied Science Publishers Ltd, London, 2nd edition, 1976. [Chapters 1–7]

Givoni, Baruch, *Climate Considerations in Building and Urban Design*, Van Nostrand Reinhold (USA), 1998. [Chapters 2–7]

Givoni, Baruch, *Passive and Low Energy Cooling of Buildings*, John Wiley & Sons Inc., New York, 1994. [Chapters 5 & 7]

Goulding, J.R. and Lewis, J.O., *Bioclimatic Architecture*, LIOR E.E.I.G., 1997. [Chapter 5]

Goulding J.R., Lewis J.O. and Steemers T.C. (eds), *Energy in Architecture: The European Passive Solar Handbook*, B.T. Batsford Ltd, 1992. [Chapters 5 & 6]

Gut, Paul and Ackerknecht, Dieter, *Climate Responsive Building: Appropriate Building Construction in Tropical and Subtropical Regions*, SKAT – Swiss Centre for Development Cooperation in Technology and Management, 1993. [Chapters 1–7]

Hyde, Richard, *Climate Responsive Buildings: A Study of Buildings in Moderate and Hot Humid Climates*, E & FN Spon, London, 2000. [Chapters 5 & 7]

Izard, Jean-Louis, *Architecture Climatique, Tome 1: Bases Physiques*, EDISUD (France), 1994. [Chapters 1–7]

Koenigsberger, O. *et al.* (ed), *Manual of Tropical Housing and Building, Part 1: Climatic Design*, Longmans, 1973. [Chapters 1–7]

Konya, Allan, *Design Primer for Hot Climates*, Architectural Press Ltd, London, 1980. [Chapters 1, 5 & 7]

Lavinge, Pierre, Chatelet, Alain and Fernandez, Pierre, *Architecture Climatique, Tome 2: Concepts et Dispositifs*, EDISUD, France 1998. [Chapters 1–7]

Lippsmeier, George, *Tropenbau: Building in the Tropics*, Callwey, Munich, 1980. [Chapters 5 & 6]

Randal Thomas, Max Fordham & Partners (ed.), *Environmental Design: An Introduction for Architects and Engineers*, E & FN Spon, London, 1996. [Chapters 2 & 7]

Titles relevant for further study: technical

CIBSE, *Natural Ventilation in Non-Domestic Buildings*, Applications Manual AM10, CIBSE, London, 1997.

Comite d'Action pour le Solaire, *Guide de l'Architecture Bioclimatique, Tome 1: Connaître la Base; Tome 2: Construire avec le Climat*, LEARNET – Comite d'Action Pour Le Solaire, Paris, 1996.

Daniels, Klaus, *The Technology of Ecological Buildings: Basic Principles and Measures, Examples and Ideas*, Birkhauser Verlag, Basel, 1997.

Daniels, Klaus, *Low-Tech Light-Tech High-Tech: Buildings in the Information Age*, Birkhauser Verlag, Basel, 2000.

Fanger, P., *Thermal Comfort*, McGraw-Hill, New York, 1973.

Institut Català d'Energia, Barcelona, *Less is More: Energy Efficient Buildings with Less Installations*, for the European Commision (DG XVII).

Olgyay, Victor, *Design with Climate: A Bioclimatic Approach to Architectural Regionalism*, Princeton University Press, USA, 1963.

Ove Arup Partnership, *Building Design For Energy Economy*, Construction Press Ltd, London, 1980.

Tassou, Savvas (ed.), *Low Energy Cooling Technologies for Buildings*, IMechE Publication, London, 1998.

Titles relevant for further studies: conceptual

Baggs, Sydney and Baggs, Joan, *The Healthy House*, HarperCollins, Australia, 1996.

Cofaith, Eoin *et al.*, Energy Research Group, University of Dublin, *The Climatic Dwelling*, James & James Ltd, London, 1996.

Cofaith, Eoin *et al.*, Energy Research Group, University of Dublin, among others, *A Green Vitruvius: Principles and Practice of Sustainable Architectural Design*, James and James Ltd, London, 1999.

Drew, Phillip, *Leaves of Iron: Glen Murcutt: Pioneer of an Australian Architectural Form*, The Law Book Company Ltd, Australia, 1985.

Farrelly, E.M., *Three Houses Glen Murcutt,* Phaidon Press Ltd, London, 1993.

Fathy, H., *Architecture for the Poor*, University of Chicago Press, 1973.

Goulding, J.R, Lewis, J.O. and Steemers, T.C. (eds), *Energy Conscious Design*, B.T. Batsford Ltd, 1992.

Herzog, Thomas (ed.), *Solar Energy in Architecture and Urban Planning*, Prestel, Munich, 1996.

Holm, Dieter, *Manual for Energy Conscious Design*, Directorate Energy for Development, Department of Minerals and Energy, South Africa, 1996.

Izard, Jean-Louis, *Architectures d'Eté*, EDISUD (France), 1993.

Lechner, Norbert, *Heating, Cooling, Lighting: Design Methods for Architects*, John Wiley & Sons, Inc., New York, 2nd edition, 2001.

Lloyd Jones, David, *Architecture and the Environment: Bioclimatic Building Design*, The Overlook Press, Peter Mayer Publ., Woodstock, New York 1998.

United Nations Centre for Human Settlements (Habitat), *National Design Handbook Prototype on Passive Solar Heat and Natural Cooling of Buildings*, UNCHS (Habitat), Nairobi 1990.

Titles on relevant specific subjects

Elawa, Sami, *Housing Design In Extreme Hot Arid Zones with Special Reference to Thermal Performance,* Department of Building Science, University of Lund, Sweden, 1981.

Etzion, Yair (ed.), *Architecture of the Extremes*, The International PLEA Organisation, Conference Proceedings, July 1994.

Facey, William, *Back to Earth: Adobe Building in Saudi Arabia*, Al-Turath & London Centre of Arabic Studies Ltd, 1997.

Fathy, H., *Vernacular Architecture: Principles and Examples with Reference to Hot Arid Climates*, University of Chicago Press, 1986.

Golany, Gideon, *Desert Planning*, Architectural Press Ltd, London, 1982.

Golany, Gideon, *Housing in Arid Lands: Design and Planning*, Architectural Press Ltd, London, 1980.

Holdsworth, Bill and Sealey, Antony F., *Healthy Buildings: A Design Primer for a Living Environment*, Longman Group (UK) Ltd, 1992.

McIntyre, D.A., *Indoor Climate*, Applied Science Publishers Ltd., London, 1990.

Norton, John, *Building with Earth: A handbook*, Intermediate Technology Publications Ltd, London, 1998.

Saini, Balwant Singh, *Building in Hot Dry Climates*, John Wiley & Sons, London, 1980.

Steele, James, *An Architecture for People: The Complete Works of Hassan Fathy*, Thames and Hudson, London, 1997.

Steele, James, *Hassan Fathy, Architectural Monographs*, Academy Editions, St Martins Press, New York, 1988.

Steele, James, *Sustainable Architecture: Principles, Paradigms and Case Studies*, McGraw-Hill, New York, 1997.

Stulz, Roland and Kiran, Mukerji, *Appropriate Building Materials: A Catalogue of Potential Solutions*, SKAT & IT Publications Ltd. (UK), 1993

Szokolay, S.V., *Architecture and Climate Change*, RAIA Education Division, Australia, 1992.

ILLUSTRATION AND DESIGN CREDITS

Photographs are by the author with the exceptions of those listed below. Th copyright holders are listed alphabetically (with page numbers and the locatio on the page - top: t, right: r, left: l, middle: m and bottom: b).
Arup Photo Library, London, pages 19 t, m + b, 95, 127 bl.
BOTEC, Botswana, pages 67 tl + bl, 136 tl.
Bidinger, Paul Erik, Denmark, page 91.
Dahl, Torben, Denmark, page 82.
Development Workshop, France, pages 48 b, 49 tl, 58 l, 75 b, 78 t, 126, 127 br.
Hansen, Martin, Denmark, page 93 br.
Kerchlango, Jörk, Denmark, pages 48 t, 51 tl, 53 b, 54 r, 55 t + b, 56 m - b, 64, 73 l, 80 t + b, 130.
Moon, Angus, photomoon.com, page 136 tl.
Murcutt, Glenn, Australia, pages 79 b, 87 t + b.
Reimann, Gregers, Denmark, page 93 mr.
Runge, Peter, Denmark, pages 54 l, 86, 94 tr, 127 tr.
Stringer, Richard, Australia, page 51 tr.
Sørensen, Peter, Denmark, pages 56 t, 63 tl, 68, 76 l + r, 77 t, 79 t, 81 + r, 94 tl.
Utzon, Jan, Denmark, pages 53 tr, 58 tr, 61, 62.
Vrdoljak, Katarina, Australia, page 93 ml.
Warming, Ida, Denmark, page 58 br.
Warming, Michael, Denmark, pages 50 b, 63 tr + br, 65 tr.

If copyright information is misleading it is hoped that no infringement ha occurred and that future editions will enable any further acknowledgement due to be gratefully made.

The architects and engineers who have designed the buildings on the photos ar mentioned in the illustration text, whenever this information was available.

The following drawings are original drawings from/by: Arup Photo Librar London of a building in Nottingham, UK, Architects: Michael Hopkins an Partners, Engineers: Ove Arup and Partners (on front cover and page 18 below and John Norton, Development Workshop, France of a traditional building i Cairo, Egypt (on back cover and page 18 top).

Line drawings, plans and tables are, with the above exceptions, produced by th author. Some tables are based on information/studies from several of the mai reference books (listed in the bibliography) and consequently the term 'variou sources' is used as a reference in the illustration text. In cases where they refer t a specific reference book or other documentation, it is mentioned in the text.

The cover design is by Hedda Bank and the author. The initial stage of the boo design was by Simon Fraser. Eddie Behrens and the author designed the fina version of the book. Eddie Behrens also digitised all the drawings appearing i the book based on sketches by the author.

Afterword

Having read this book, you are probably aware that active design is not that easy. If you want to do something a little easier, you could always try making a 'Norwegian Ice' or 'Baked Alaska'. Not only is this a delicious dessert, but it also provides a very convincing model of thermal thinking and in particular insulative properties!

For six people, you will need: one litre of ice cream (vanilla or any flavour of your choice), a thin sponge or angel cake (about one centimetre thick), the whites of approximately 8 eggs (depending on size) and a few spoonfuls of sugar. You may also like to add marzipan, crushed almonds or macaroons, or a liqueur such as brandy or cognac.

Begin by whipping the egg whites with a couple of spoonfuls of sugar until the mix is stiff. Put the sponge or angel cake in a cake tin (you could also use crushed macaroons - the base just needs to be an airy, insulating substance) and add any extra ingredients (marzipan, almonds, liqueur etc). Cut the ice cream into pieces and place it over the cake (it should not spread over the edges). Then cover the ice cream with the whipped egg whites. The layer should be about 2 cm thick. Sprinkle a final spoonful of sugar over the top and then place the cake in a 200–250°C preheated oven for 5–8 minutes, i.e. until the sugar lightly browns.

For the finishing touch, just before serving, pour some cognac or rum over the dessert and put a match to it!

And there you have it – a literally blazing hot topping to a freezing cold inner layer. What better example of staying cool?

Good luck!
holger@koch-nielsen.dk

The author warmly welcomes readers' comments and any suggestions they may have for additional or alternative material, including photographs or illustrations, which might be incorporated – suitably credited of course – into any future editions of this Guide.

INDEX

S

T

U

V

W

Z